DOLLS THE WIDE WORLD OVER

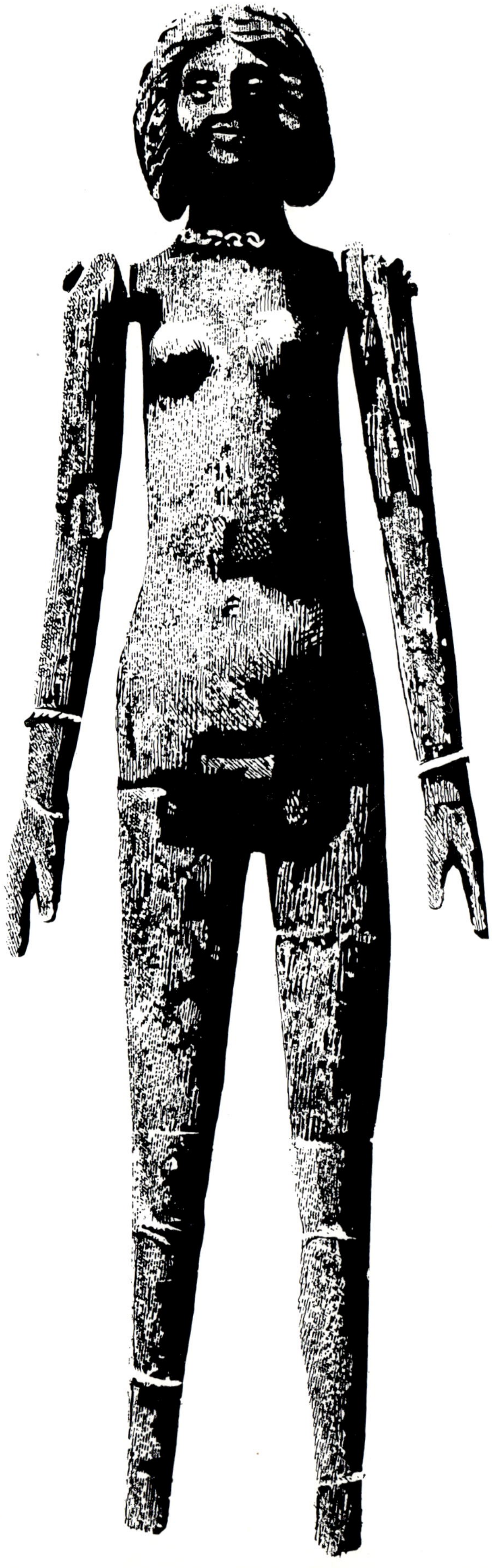

DOLLS

THE WIDE WORLD OVER

by Manfred Bachmann and Claus Hansmann

translated by

Ruth Michaelis – Jena

with the collaboration of

Patrick Murray MBE, FSA (SCOT)
Curator of the Museum of Childhood, Edinburgh

CRESCENT BOOKS NEW YORK

Text: Manfred Bachmann · Photographs, drawings, and design: Claus Hansmann

Dustjacket, front: Wax doll in christening-robe with movable arms and real hair, England 1880
Dustjacket, back: Embroidered miniature of Christine von Bayern as child, with doll, 1576
Page 4: Late 19th-century wax doll, presented by Queen Mary to the Bethnal Green Museum, London
Page 6: Dolls made from twig forks, Bissagos Islands, Guinea-Bissau; the doll on the left is from the Carraya tribe, Brazil

Library of Congress Catalog Card Number: 72-94617
 This edition is published by Crescent Books a division of Crown Publishers, Inc.
a b c d e f g h
Manufactured in the German Democratic Republic

ISBN: 0-517-181762

Contents

Introduction

Play activates body and mind. It enriches the imagination and feeds the emotions. Even in these days of rapid technical progress play remains essential to the child and, with certain qualifications, to the adult also. In his writings the German philosopher Georg Klaus maintains that no grown-up man need be ashamed of admitting to his love of play, that, indeed, many a career was chosen because work in a particular field happened to correspond to something played at in childhood. Play, says Klaus, far from being a waste of time, has a deeper meaning, and increasing attention should be paid to it in the training and education of the young. Experience has shown that crammed lessons are less effective than learning through playing when the child absorbs knowledge spontaneously. Such knowledge is lasting.

Play generally implies not only toys but a wide range of things from simple objects turned haphazardly into playthings to those made specifically as toys. The doll, a truly representative toy, should help in establishing a child's proper relationships to man and animal. The design of dolls is subject to the same rules as that of all good toys: it must give the child scope to realize his own ideas. The proper balance of play possibilities and play desires alike is important in the choice of the thing to be played with, for it should stimulate the child to creative mental and physical activity. Play thus becomes a step forward on the path of learning, the final aim being mastery of the whole art of living. Accessories to the main plaything are also required; in the case of dolls, this means prams, furniture, additional sets of clothes and so on. Handling all these things effectively will also teach the child a sense of order.

Although toys must be both functional and aesthetically satisfying, it could be said that the very perfection of some popular modern dolls, equipped with mechanical and acoustic devices as they are, endangers their value as playthings. Their potential is obviously limited if a child cannot dress or handle them with freedom and affection. This point is doubly important since unrestricted and imaginative play with dolls can greatly influence the child's relationships with his parents, brothers, and sisters. Also in this way souvenir dolls with a difference of colour or costume can help to cement friendship and sympathy with other peoples.

Finally, all toys should be suited to the particular child's age and comply with its intellectual demands. Thus, the older and more developed the child, the bigger, more delicate, and more sophisticated must be the doll. This apparent contradiction of the earlier statement about dolls of this type vanishes if one accepts that the sophistication of the toy or doll must obviously match that of the child. Friedrich Froebel's statement about

1 *Soft dolls, from a child's grave. Native cloth. Peru, c. 1600*
2 *Doll's house, belonging to the Nuremberg patrician family Stromer. 1639*
3 *Doll's house (bureau type), of Petronella Brandt, née Ortmann. From an Amsterdam workshop, c. 1700*
4 *The king's bedroom. From the doll's house at Windsor Castle, belonging to the Royal Family. 19th century*
5 *The painter in his studio. Oil painting by Adriaen van Ostade. 1663*

play in general will also serve specifically for the play with dolls:

> Play, then, is the highest expression of human development in childhood, for it alone is the free expression of what is in the child's soul. It is the purest and most spiritual product of the child, and at the same time it is a type and copy of human life at all stages and in all relations. So it induces joy, freedom, contentment, inner and outer repose, peace with all the world. From it flows all good.

Another interesting aspect of the doll, showing how closely related it is to human activities, is its increasing use in psychotherapy. Through playing with dolls the young patients recreate the environment and social climate in which they wish to live. Repressed emotions and impulses are released as, for example, aggressive feelings towards brothers and sisters. Sylvia Cassels had remarkable success with her doll therapy in the children's ward of the Wesley Memorial Hospital in Chicago, as is shown by a report of 1968 which states that in application of her theory she used dolls to represent patients, nurses, anaesthetists and surgeons. Boys and girls watched a 'patient' doll having her arm wiped with an ether-soaked piece of cotton wool before having an injection. In this way the whole procedure of the operating theatre was experienced as a form of play. The children were in consequence found to have lost their dread of the unknown, and those who had had this preparation in play for their eventual treatment recovered more quickly, and cheerfully returned to the hospital for after-care.

This book is mainly concerned with the doll as a toy, and does not go too deeply into its almost numberless other aspects, most of which elevate it to the status of a figurine rather than a plaything, and find it highly charged with social, ethnic or didactic implications. Nor does it treat any of the vast related fields in which the real social significance of the doll is first hinted at. It would require something more than an encyclopaedia to deal adequately with dolls' rooms, kitchens, shops and houses, with dolls' costumes, furniture, and toiletry, to say nothing of dolls' prams, crockery and even jewellery. Numerous writers have dealt with most of these subjects, though definitive books have yet to be written about any of them.

The general approach is historical, taking into account the educational and psychological aspects of the doll, and following the lines of an earlier volume on toys. (Karl Ewald Fritzsch and Manfred Bachmann: *An Illustrated History of Toys*. London, 1966.) It does not claim to be a full historical account, as no single book could do the subject justice, particularly in the ethnographical field, and the authors are only too well aware

KING'S BEDROOM

6 *Living room with lady and child. Oil painting by Johann Heinrich Wilhelm Tischbein. 1777*

of the danger of identifying as toys much of the early grave material that has recently come to light.

In the first chapter will be found a brief mention of the religious use of dolls and what von Boehn calls the "first groping towards plastic form" in the early history of mankind. There follows a discussion on oecological types, made from materials found naturally in the child's surroundings, and so on to dolls produced by craftsmen and industry.

In the third chapter the doll, both that of the rich child and of the poor, is shown as a mirror of contemporary social life, examples being taken from different centuries. Then the specific doll cultures and traditions of East and South-East Asia are described. The last chapter deals with the modern doll, which is tending to become a showpiece rather than a toy, something which can be attributed to the increasing interest in souvenir dolls.

The remaining chapters deal with the historical aspects.

A word must be said about the lay figure, the large or small 'doll' used by artists (Nos. 5 and 173). It was known as early as the fifteenth century, but in our day the most notable example is that of Oskar Kokoschka. In 1918–19 he had a life-size figure made for him by a Stuttgart craftswoman, and eventually it was to accompany him everywhere. He used it as a basis for some 160 studies in various techniques, rendering her lying, sitting, and standing, dressed and nude. In 1919 he produced from it his well-known *Woman in Blue*.

In modern art, with its often intentional destruction of the values of the human image, the doll is used repeatedly to suggest the impermanency of normal life, and often these interpretations have sadistic or pornographic implications. In 1968 the painter Manfred Schmale exhibited what he called *Objekte* in a West Berlin Gallery. These were unpleasant arrangements of bruised and chopped up doll bodies, and typical of his approach is a wretched doll-like object lying on a plate, next to which are the necessary knife and fork. The implication is obvious.

In other examples doll bodies are shown forced into preserving jars, or with limbs dislocated, burned, mutilated, drenched in acids or shrunk and wrinkled by soaking in turpentine. In this mood Eva Aeppli, who lives in Paris, gave an exhibition in Nuremberg in 1969–70 consisting of life-size figures made of various textiles. These were grouped in curious arrangements, their poses suggestive of a wraith-like existence close to death; notable among them were *Four Widows* and *Verbena in a Wheelchair*.

In a different medium a Frenchwoman, Niki de Saint-Phalle, produced at the beginning of the present 'pop' era large plaster and synthetic figures, which she called

7 *Doll. Palm-leaf stem (triangular in section), wax breasts, belt of plant fibre, dyed yellow. Canella tribe, East Brazil*

Nanas, and which could be manipulated into various and suggestive poses. On a very much larger scale was a female doll twenty-seven metres long lying on her back, which Niki de Saint-Phalle showed in 1966 at Stockholm. It proved a sensation, for it was hollow, and was visited by no fewer than 2000 people a day, the entrance being sufficiently obvious. The doll was described as "a gigantic female cathedral"; it might be thought of more as evidence of a morbidly inclined world.

Yet all these fringe manifestations must not divert us from our real object: the translation of the doll from its initial simplicity as a child's toy to the luxurious models of today in all their technical perfection. This progression is common to all ages, each of which produced something particular to itself—both good and bad.

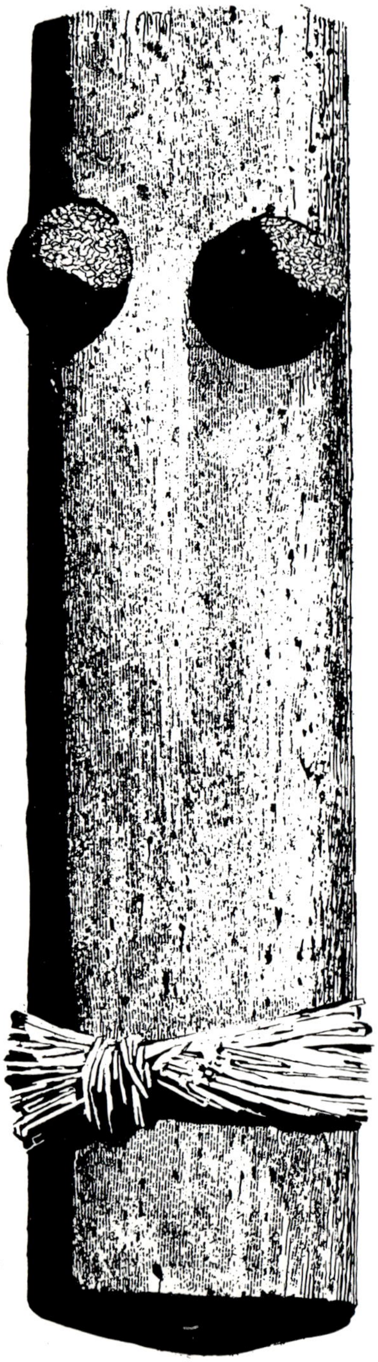

7

I Primitive Forms of the Doll

A glance at the world of today's child shows that the old idea that an object has to prove its worth as a toy still holds good, and that the toy fulfils this function only when it is accepted by the child as a plaything. Even a shop-window figurine may become a toy if it gets into the hands of a child who regards it as a doll, and takes it into its confidence as a playmate. Educationists, sociologists and ethnologists all agree that play is basic to the child's coming to terms with his surroundings. Playing with dolls is a preparation for life; it is playing a part, anticipation and imitation at one and the same time. It is closely related to real life, and it is typical of this kind of play that the child imitates the acts of adults, of mothers and children. This does not mean that the doll must be the faithful replica of a child, for when there is no proper doll any wooden figure or even a piece of wood will suffice to make a girl play. In playing with dolls she performs a specifically human act, as no young animal plays with dolls or has the need to do so. The girl regards a doll as her real child.

The historical age of the doll as a toy cannot be determined with certainty. Recent research suggests that definite proof of the doll serving as a toy cannot be found before the late Egyptian period. In prehistoric times doll-like figures were connected with cults and religions as idols, fetishes, and amulets. This is known through grave-finds and through ethnographical research among peoples in remote areas with a still primitive way of life. We may remember here the many small dolls found by Schliemann in Troy which are now recognized as ancestor figures. Schliemann himself took them for idols. In more modern times there were the little Chinese dolls made of soapstone—the best belonging to the sixteenth to the eighteenth century—which stood on small altars as guardians of the home. On festive occasions these were given as luck-bringing gifts, thus pointing to the transition from idol to amulet.

This short summary concerning the ancestry of the toy doll does not support earlier writers on art who see the origin of all plastic art in religion. The American ethnologist Ruth Benedict has this to say in describing the results of her extensive field work among the Pueblo Indians of New Mexico:

> As a matter of history great developments in art have often been remarkably separate from religious motivation and use. Art may be kept definitely apart from religion even where both are highly developed. In the pueblos of the south-west of the United States, art-forms in pottery and textiles command the respect of the artist in any culture, but their sacred bowls carried by the priests or set out on the altars are shoddy and the decorations crude and unstylized ... "We have to put a frog there," the Zuñi Indians say,

8 Doll. Animal bone, with bead ornamentation. Central Africa

9 Dolls. Bone, with wax breasts, glass bead ornaments, eyes of red fruit kernels. Togo

10 Dolls. Ivory, incised ornamentation, movable arms; holes in heads for fastening wigs. Coptic, Egypt, 2nd–4th century A.D.

11 Doll. Ivory, black painted, incised ornamentation, wool dress. Coptic, Egypt, 7th (?) century A.D.

8

9

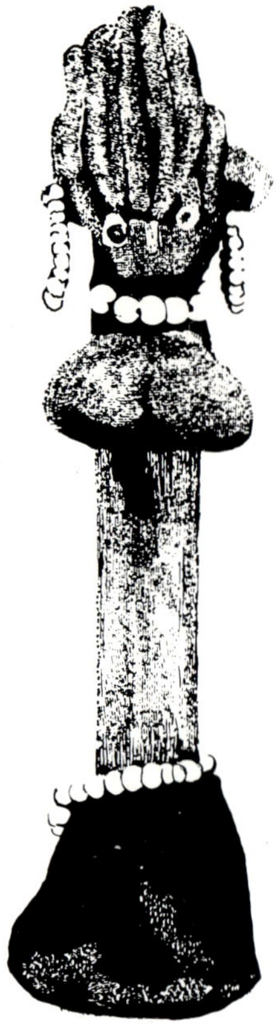

meaning that the religious exigencies eliminate any need of artistry.

On historical principles it seems unlikely that art evolved just on the basis of a desire for creative achievement. It had to have a dynamic, positive, even if in itself less independent driving force, which would give it at once both the necessary stimulation and impetus. Obviously, if this theory is correct, religion would do all this, and do it best in the earliest art of all, that of the caves which itself drew its inspiration from the earliest religion of all, sympathetic magic.

Ethnologists now believe that in the early phases of cultural development not every artistic manifestation had a definite purpose within a definite sphere of life. Everywhere and at all times there may have existed the need for man to express his creative faculties, if only in play, as enjoyment, recreation, or satisfaction of the senses.

It has been said that if we recognize work as the source of play, then the concept of work must not be taken too narrowly, and certainly it must not be considered as limited to man's relation with his immediate surroundings. It has rather to be seen against the backdrop of the whole world, and as something which, however remotely, is going to affect the life of the whole community. It also has intimate effects on the individual, conditioning the relationship of children to their own and other groups, their conduct towards their parents, the old, and members of the opposite sex. Ideas of good and bad arise in play, the permissible and the antisocial, in short, most of the incredibly complex code which in all societies has governed the behaviour of man towards his fellow creatures.

Children's play is inevitably going to colour their outlook, strengthening the good and the bad equally. Thus their games, apparently trivial, are far from being

10

11

so. While on the surface only innocent reproductions of adult working activities, they are in fact true railheads of future lines of conduct. However difficult it may be to determine just what vitally affects each individual child, there can be no doubt that children very early show strong social sense, whether by group activities or by strictly personal relations, say of a girl with her doll. Acting the little mother to it is an entirely new human experience, but it is also an excellent display of accepted conduct since it entails playing a part and acting according to rules.

It is uncertain whether simple dolls to play with existed in earliest times, side by side with dolls serving religious or cult purposes, since none have survived the millenniums. The fact that some peoples who have remained in primitive conditions do not know the doll as plaything, or have come to know it rather late, is not a reliable guide. According to American ethnologists, the Indians took over dolls for playing from immigrants only at the time of Queen Elizabeth I. Some Red Indian languages have no word for doll, and in others it is identical with 'baby'.

Between the doll for playing with and the cult doll stands the vividly painted *kachina* doll of the Hopi and Zuñi, the Pueblo Indians of the south-west of North America. *Kachinas* belong to the ritual of a secret society to which all male adults of the tribe belong, and to which women can be admitted in special circumstances. The organization has several subdivisions, each with a separate *kiva*, a subterranean room for ceremonies, its own spiritual leaders, dancers, and body of members. The central figures of the cult are the masked spirits, the *kachinas*. These spirits are regarded as the mystical inhabitants of the bottom of the lake, lords over cloud and rain, creators of fortune and wealth. They are kindly disposed, friendly gods, fond of dancing, guar-

anteeing the fertility of fields, causing rain, giving blessings of many kinds, protecting and training children. Ruth Benedict has studied the Zuñi ceremonies in detail, and has described them thus:

> There are more than a hundred different masked gods of the Zuñi pantheon, and many of these are dance groups that come in sets, thirty or forty of a kind. Others come in sets of six... Each of these gods has individual details of costuming, an individual mask, an individual place in the hierarchy of the gods, myths that recount his doings, and ceremonies during which he is expected.

The masks used in these ceremonies are sacred objects. A man putting on the mask of a certain god becomes that god himself. Of the different *kachina* masks, some are particularly meant for gods that punish, and they frighten naughty children at the annual festival. The first initiation of children into the secret cults takes place between the ages of six and ten by a confrontation with masked dancers acting as the *kachina* spirits. The same *kachinas* are represented also by wooden dolls, called *tiku*, which serve for the children's instruction and surrogate for the spirits. They are used as toys in a secondary function.

The painted decoration of the *kachina* dolls follows the exact colouring of the masks and is an important

12 *Doll. Clay, probably based on older bone dolls; red feather ornamentation. Caraya tribe, Brazil*

means of identification, showing where the dolls come from.

The transition of the doll from its function in religion and cult to being a toy, as seen in the example of the *kachina*, is not the only way in which once sacred objects can get into children's hands; one example out of innumerable ones is that mentioned in 1544 by Roger Edgeworth, Canon of Bristol Cathedral, who complained that Catholic churches were being plundered and figures of the saints and Our Lady given to children as dolls. In Japan little girls do not throw away their worn-out dolls, but often dedicate them to some deity at whose shrine these doll remains come to rest.

Ethnology so far has paid but little attention to toys. There have been few if any systematic investigations, and then their findings consist more of theories than of solid conclusions. Particularly is this true of the doll, though most of what is now considered 'modern' in the thinking about it was realized over ninety years ago by an Englishman, Edward Lovett. Unfortunately, the isolated mention of dolls to be found in reports of expeditions gives no true picture of their status in an undeveloped society. This is the greater pity since it is only in such societies that the true origin and social significance of the doll can best be studied. Max von Boehn may have oversimplified in his statement that the doll has no real antiquity.

Over the question of the materials used certain general conclusions can be reached.

Among primitive peoples children's dolls were, and usually still are, made of the simplest materials. In the lower Congo girls are reported to have had tied to their backs a piece of wood or cassava root which they called *bena*, child, and in that way they played at carrying their own children. They showed no interest in the European dolls that were occasionally given to them, so that their mothers soon took these dolls from them and sold them as fetishes. A girl of the Crao Indians in South America is reported to have played with two empty corn cobs which she lovingly nursed as her real children, while doll-like clay figures made by the Caraya Indians, reflecting the life of the tribe, were not accepted as toys by the children.

For many peoples bones of a suitable shape and hardness formed the raw material for many utensils such as fish-hooks and arrow-heads. Dolls, too, were made of bone. Two dolls of the Haussa of Central Togo have bodies made of bone, while the breasts, stuck on, are of beeswax dyed black. Necklaces, belts of beads and other ornaments underline their female sex (No. 9). With other dolls of the same type hair-styles have been carefully arranged, and for a life-like representation of pubic hair, human hair is used. Bone dolls are common all over Africa. The German ethnologist Paul Germann

13 Doll. Wood. Island of Usedom, German Baltic coast, 11th century
14 Dolls. Wood with pokerwork. Mossi tribe, Africa
15 Doll. Wood, linen dress. Africa

believes that they have undergone the transition from sacrificial doll to toy, and quotes Herbert Bastian on a study of the Yoruba in present-day Western Nigeria:

> At certain times they have a sacrifice for the dead when a householder kills a ram, letting the blood run into a hole in the ground. The family meal ended, one of the bones is dressed, preferably in the suit of a white man. Then the bone is carried about the town as the bone of an ancestor, while women walk in front of it, chanting its praises.

Several collections also possess early forms of dolls made of ivory (Nos. 10, 11, 30), the surface of which lends itself well to carving and ornamenting (engraving, colouring or painting).

A clay doll from South America, greatly resembling bone dolls (No. 12), illustrates the basic factor limiting the creative activities of primitive peoples—that is, climate and environment determine the raw materials available. This does not necessarily mean that the craftsman will let the material, which may either stimulate or hinder him, subdue his imagination. Often form, style, and manner of expression remain the same even though realized in different materials.

Bone carving is still one of the most important techniques in a number of civilizations. In Tobolsk in Western Siberia miniature figures, often souvenir dolls, are made of mammoth- and walrus-bone. Their theme is the working life of the one-time nomadic Ostyaks, their hunting and fishing. Bone carving is documented for 1500 years in the arctic north of the Soviet Union. It is still practised by the inhabitants of northern Siberia, the Evenky, the Chukchi, and the Nentsy.

The Eskimos show great skill in bone carving. These people of the North, who are few in number, and who live in the meagre coastal regions of the U.S.A., Canada, Greenland, and the U.S.S.R., in a hard climate, work bone in many different ways. In 1936 Hans Himmelheber visited the Eskimos on the Kuskokwim river in Alaska, studying their ancient culture. He found that the only artistic creations among their toys were dolls. The heads were most carefully worked in wood or ivory, and the hands were of the same material or of hide. In northern Canada, in the midst of the ice and snow of Hudson Bay, some three hundred Eskimos of Cape Dorset, a small island on the same latitude as Baffin Island, are known specially for their skill. Their expressive sculptures of stone and ivory as well as their drawings on bone or stone, reflecting the Eskimo way of life, can be seen in many museums and exhibitions in Europe and America.

Wood is one of the oldest raw materials which man has used freely for his creative work. The great significance of wood lies in its technical qualities. It is comparatively easily worked, it can be used in many different

13

14

15

16 *Doll. Wood, coloured thongs, beads, and cloth dress. Africa*
17 *Doll. Wood, roughly carved, remains of ochre colouring. Nupe province, Nigeria*
18 *Dolls,* kachina. *Wood. Hopi tribe, New Mexico*

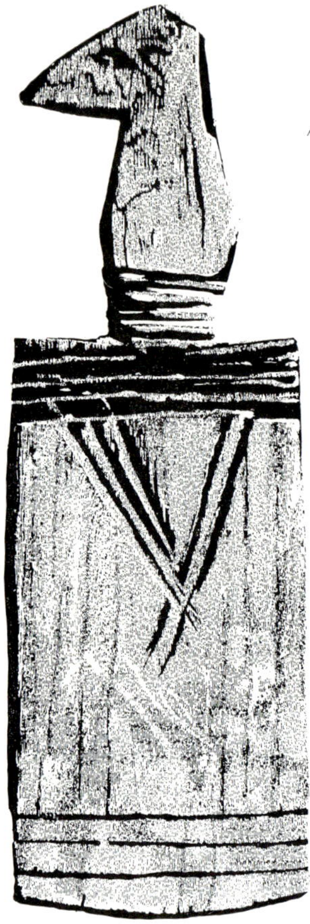
17

18

16

ways, and it is usually readily available. Its colour, marking and texture make it particularly suitable for artistic use.

Among the African wooden dolls in the Leipzig Museum für Völkerkunde a number of little figures, about some 15–20 cm high, particularly attract attention. Their trunks consist of a wooden cylinder broadening at the base, suggesting feet. There are no legs, and arms are only hinted at. Breasts and navel are shown as tiny humps, while head and face are merely outlined (Nos. 14, 15). Similar dolls are mentioned repeatedly by explorers. They have a dual function: they serve as toys for little girls, and are later carried about by adolescent virgins and young childless wives who hope for a child of their own. It is reported that among the Swahili, the main population of the coastal regions of Kenya, Tanzania, and the islands lying along the mainland, young girls receive about a year after their first men-

19

19 Doll representing a youth. Gourd, head ozokerite (mineral paraffin wax). Caraya tribe, Brazil

struation a wooden doll at a great dancing ceremony. This doll they must keep with them while cooking or doing other domestic work, to make sure that they will bear children in marriage. Similar customs are known with other ethnic groups in the whole of Africa. The dolls are taken to be charms against barrenness.

In their essential form these African dolls are similar to a German doll found on the island of Usedom (No. 13).

Dolls from Ghana, too, with their outsize flat, disc-shaped heads show purposely stressed sex attributes. They wear bead ornaments (No. 31). Often these wooden dolls are only sparsely carved, and have therefore a very stylized appearance (No. 24). Even with dolls made from the simplest available materials there is a tendency to give them a life-like human appearance. It is easy to make dolls from gourds of the type used also for flasks. Through tying the plant in while growing the necessary bottle shape is achieved (No. 19). This also represents the doll's body, and by painting, engraving, and other techniques the doll can be successfully decorated. Barundi girls, the children of East African veldt farmers, use the purplish flower of the banana tree or a bottle-shaped gourd for dolls. They dress the banana stem with fringes of raffia, and make holes in the round part of the gourd for eyes, nose, mouth and ears, a few incisions on the lower part representing hands and arms. Two dumb-bell shaped gourd dolls from Mozambique have buttons for eyes, and their hairstyles are suggested by incisions (No. 29). Often they are tightly enveloped in strings of coloured beads, fastened with wax, or they may be decorated with shells. With other African dolls the body is made from a stick with two small gourds stuck at the ends, and is covered with material. Often the lower gourd has scratched into it the tattoo mark of the tribe and the attributes of sex.

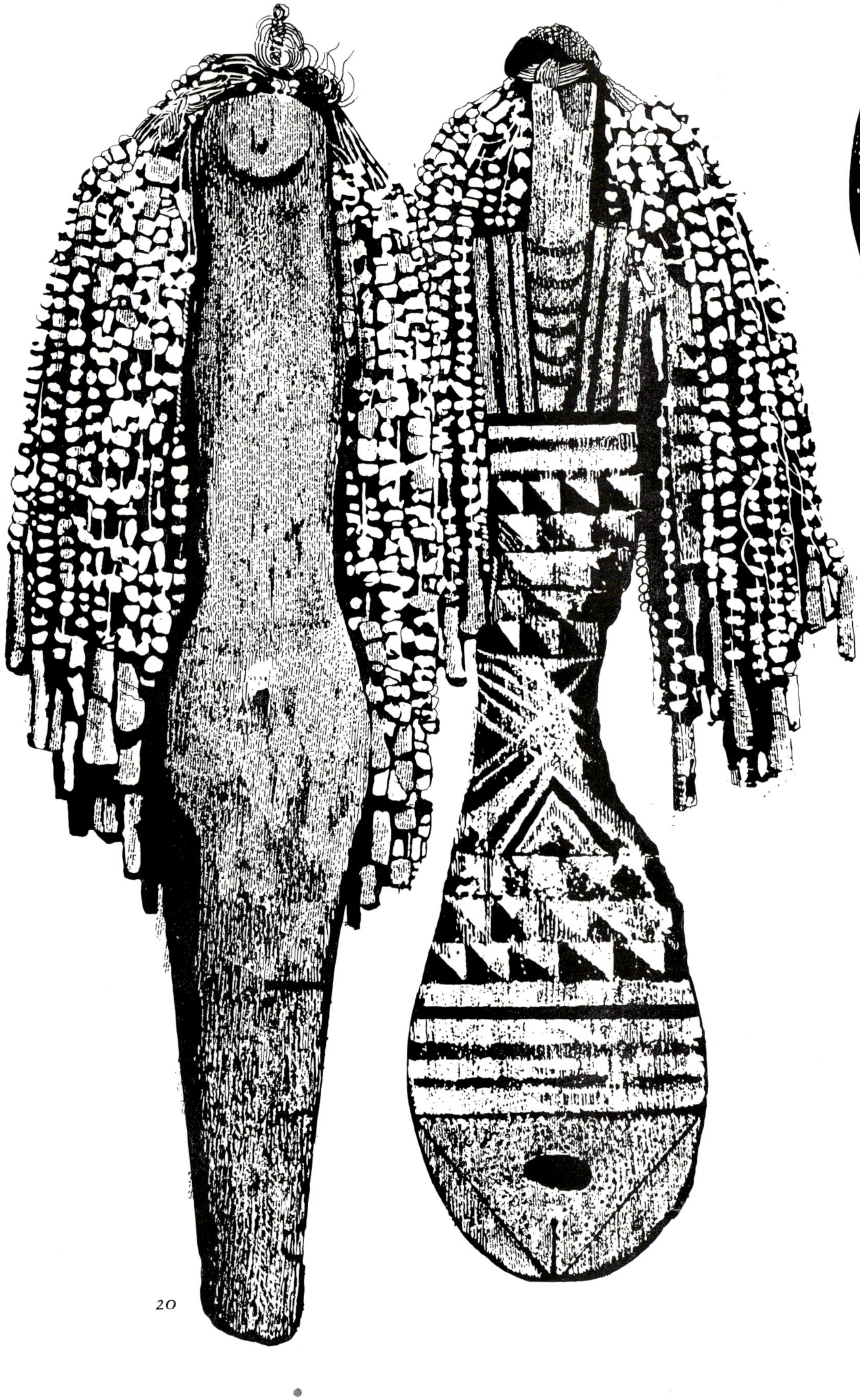

20

21

22

20 Dolls. Wood, hair made of mud beads. One with coloured ornamentation. Egypt, c. 2000 B.C.
21 Flat doll. Ebony, carved and painted, glass beads. Ghana
22 Doll. Corncob. Mexico

The women of the Wanat-Zuri, when they have given birth to their first child, are presented by the other women with a gourd enveloped in strings of beads. This establishes their status as mother. With other tribes, too, gourd dolls are symbols of fertility. It is reported about the Wa-Zaromo that if a woman cannot have children or if they die, she will buy a 'gourd child' to play with as with a real child. If eventually she gives birth to a child of her own, she must not discard the doll, but must carry it about with her like a twin. If she has a daughter the girl will play with the doll, and keep it until she herself has children. Should anybody else want to play with the doll, he or she must pay money for doing so. If the doll is put away or sold, mother and child are likely to fall ill and die. Similar customs persist among many African tribes.

Some African dolls have a corn cob for a body. Their hairstyles of carefully plaited corn straw look natural

23 *Doll. Clay, painted, linen wrap. Coptic, Egypt, 4th century* A.D.

24 *Dolls, male and female. Clay, eyes inset pieces of glass, hair indicated by incisions. Keta, Ghana*

25 *Doll. Clay, bark loin-cloth, resin-modelled hair. Caraya tribe, Brazil*

26 *Doll. Clay, painted. Peru*

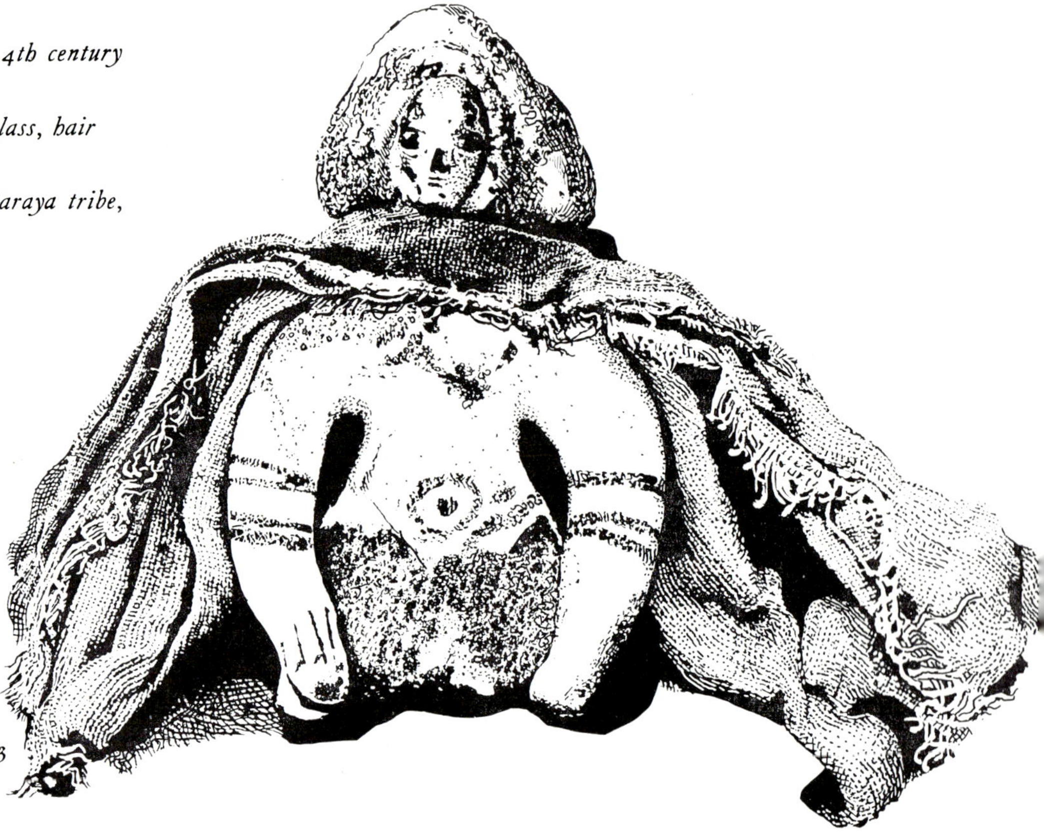
23

and are often supplemented by real hair and adorned with beads. A doll from Angola, with a corn-cob body, breasts made of wax, and dressed in bits of rags, is clearly recognizable as a woman. Mexican dolls are made of several cobs with dresses of corn straw and plaited straw arms (No. 22).

Raffia, the fibrous part of tree bark and plant stems, strips of bamboo, palm fibre, straw, even grass and leaves, are used in the making of dolls (Nos. 27, 28, 34).

In the Malayan archipelago dolls are made of strips of palm leaves, plaited diagonally and ornamented with beads. The Nentsy use the beaks of ducks, leather, or wood to make dolls' bodies, which they cover with cloth. In northern regions (among the Chukchi, in Alaska, in East Greenland, on the north-west coast of the United States) the doll made of hide is common. It is dressed like the adults of the region. Spanish influence can be traced in the dolls, which are worked in crochet or knitted with coloured wools over a wire frame. Such dolls are still offered for sale in Guatemala, Bolivia, and Chile. In ancient Peru doll-making was highly developed before the advent of the Spaniards. Grave dolls—solid gold for women, silver for men—have been recovered. They were dressed with material originally. There were also less precious dolls in Peru made of wood, wool, or straw.

Clay is an easily moulded material. With the exception of Australia and Polynesia, and a few isolated ethnic groups, pottery is known all over the world. Wherever the potter's wheel was not common, pottery was chiefly the domain of the woman. It therefore seems likely that dolls made of clay or loam were more general than museum pieces appear to suggest. Clay dolls from Africa are crudely shaped human figures. They show female sex attributes, and in some cases male ones, moulded or scratched into the clay. These dolls are very stylized, with heads and faces indicated by incisions (No. 24).

24

The ancient cultures of Mexico and Peru have left interesting examples of clay dolls (No. 26). Up to the present day ceramic 'folk art' dolls are made in these countries by craftsmen. The clay dolls of Brazilian Indians represent voluptuous women, and also reflect the life and work of the tribes (Nos. 25, 31).

Clay rattle-dolls are known in Colombia. In several parts of Africa, particularly in the north, much attention is given to complicated dolls' clothing. Their bodies are often made of reed or bamboo, and then most carefully dressed with cotton fabric, sometimes patterned (Nos. 36, 37, 38, 40). Other dolls are made of crosswise fixed sticks (No. 39). There are isolated examples of stuffed dolls made out of leather by the Zulu, or of finely plaited palm fibre filled with sand by the Swahili. Indian, Arabic, and possibly also European influences can be traced in these African-dressed dolls.

The use of simple raw materials for the making of

25

26

27 *Stick dolls. Heads ozokerite (mineral paraffin wax), raffia decoration. Caraya tribe, Brazil*

28 *Doll. Corn straw, painted face. Thai. Possibly mission work, as only wooden dolls are known from Thailand*

29 *Dolls. Gourds, pokerwork, eyes made of buttons, cloth and bead ornamentation. Mozambique, East Africa*

30 *Doll. Ivory, cotton dress with ornamental border, blackish-brown hair. Coptic, Egypt, 2nd–4th century* A.D.

31 *Dolls in a boat. Clay, decorated in yellow and black. The painting on the heads represents feather wheels, and on the bodies the tattooing of the Caraya Indians. Caraya tribe, Brazil*

27 28

dolls is also known in European folk art, particularly among mountain peasants, herdsmen, and woodmen. Fine examples are to be found in the south-east of Europe, in Switzerland, Sweden, and Austria (Nos. 42, 68). In present-day arts and crafts this type of doll has become fashionable, but it has no connection with genuine tradition. A distinction must be made, however, between conscious 'primitiveness' and the serious attempts of well-known artists to take tradition and folk art as a source and inspiration for their own creative work. Working with wood, for example, was given fresh direction by Arthur T. Winde of Dresden, whose followers in the twenties and thirties greatly influenced traditional wood crafts and industry in the Erzgebirge through outstanding aesthetic developments (No. 41).

It is natural that every mother should want to make dolls for her own child. These dolls can be made of anything—wool, fur, and all kinds of bits and pieces, felt, wire, and even old stockings. In Bavaria such a doll is known as *Fleckeldocke*, meaning a doll made of scraps. Dolls made of oddments are mentioned in *Arche Noah*, published in Dillingen in Germany in 1693. It is also likely that this type of doll was known in the early days of history, and among primitive peoples. Its very perishable nature appears to be the reason why hardly any examples have come down to us, though the type is known all over the globe. For example, a report from Transylvania dating from about 1915 states:

> Dolls are very common with our peasants. A wooden cooking spoon often forms the base on which the clothes are arranged. The head is covered with a scrap of white material, and cheeks and lips are painted on it with the juice of beetroot, while nose and eyes are drawn with coal.

34

35

32 *Twin doll. Wood, bead ornamentation. Haussa tribe, West Africa*

33 *Doll. Wood, carved and painted. Nigeria, present day*

34 *Doll. Raffia, balled head. Brazil*

35 *Doll, with birds. Peru*

36

37

38

39

40

36 Native dolls. North Africa

37 Doll. Millet stalk, bead ornamentation, cloth wrap. Zanzibar, East Africa

38 Soft doll. Body and dress of different materials. Barbary, North Africa

39 Doll. Contrived from a wooden stand. Africa

40 Soft doll. Coloured dress, bead ornaments. North Africa

41 *Flat dolls. Wood, painted faces. Reinhold Langner, German Democratic Republic, mid-20th century*

42 *Doll. Straw, torso dark blue raffia. Austria, mid-20th century*

41

42

2 *Craftsmen-made Dolls*

The oldest wooden dolls made by craftsmen can be dated roughly 2,000 B.C. and come from Egypt. Cut from a thin piece of wood, their bodies are covered with a geometrical design suggesting clothes. Wigs made of strings of wooden or clay beads, simulating hair, are fixed to the dolls' heads with natural paraffin wax (No. 20). The arms and legs are often movable. The earliest surviving stuffed dolls were discovered in the Coptic graves of Ekhmîm Panopolis in Upper Egypt and they date from the sixth and seventh centuries A.D. Some are made simply of coloured wool, others have painted wooden heads and coloured woollen clothing. There are also dolls with bodies made of alabaster, ivory, or clay (Nos. 10, 11, 23, 30). Frequently the dolls' clothing is modelled on that of the adults of the period. In the place where these dolls were found, Robert Forrer came upon the tunic of a child of three to three and a half years, which had been decorated with blue dye by the process of wax-painting, and is said to be the earliest example of resist or batik printing (fourth century A.D.). Experts now agree that in the late antique art of Egypt objects made by skilled Coptic craftsmen, including dolls, were true folk art, reflecting the aesthetic demands of the time.

In Greece the colloquial word for doll is *koré*, meaning girl. Making dolls was obviously one of the main tasks of the *kora-plathos*, the sculptor. The word *koreia* described at the same time small clay objects, gifts dedicated to the gods, and this may indicate the connection between toy and sacrificial gift. Most Greek dolls were made of burnt clay (terra cotta), their limbs fixed with string, wire, or gut. There were also dolls made of wood, wax, gypsum, and ivory. In the eighth and seventh centuries B.C., and with the growing production of pottery, there developed a large home and export trade in Greek vessels, particularly from Corinth. It may be assumed that dolls were included in this trade. In the fifth and fourth centuries B.C. a greater specialization of traders came about, and the producers of terra-cotta objects, toys and sacrificial gifts of clay or wax, separated from the potters. In that way Greece had its first toy-making 'industry'. Sardis, the capital of Lydia, favourably situated on a trade route to Persia, became a famous centre for toys.

The oldest known Greek dolls came from Boeotia. They have movable legs under bell-shaped skirts. Their dress is richly ornamented with geometrical patterns, swans, fish, birds, and dancers. Common all over the Greek world were simple jointed dolls (Nos. 44, 47) representing adolescent girls. Stiff arms and legs hang on wires pushed through the shoulders and hip joints. Girls played with their dolls until they were married. Then, in a solemn ceremony, they dedicated the dolls with all their clothing to Hera, the protectress of

43 Dolls. Wood, painted. Korea

44 Doll. Linen dress, natural hair, painted eyes; necklace, anklets, and bracelets of leather with gold foil. Egypt (Greek source)

marriage, to Artemis, the goddess of chastity, to the nymphs, or to Aphrodite, the goddess of love. There is a Greek epigram telling of a girl, Timarete, who dedicated her dolls to Artemis, together with all the doll's clothing, a tambourine, a favourite ball, and her own little bonnet.

Dolls stood for a wish fulfilment. Grave reliefs of the fourth century show pretty young girls charmingly engaged with their playmates of clay, indicating that these girls died before marriage, and unfulfilled. Attic dolls represent a special type with stumps for arms and legs. In their rigidity they are reminiscent of carved images, and are thought to belong to the time of the Persian Wars. The flat Corinthian jointed dolls are from the late Archaic period, as are types of sitting dolls in Attica, Boeotia, and the Greek colonies. Among these sitting dolls are representations of naked boys. The great beauty of Greek dolls seems to indicate that they did not serve a practical purpose only—the preparation of the girl for her future role as mother—but that they were to show at the same time the human ideal.

Roman dolls were mostly made of clay, painted sometimes, and very popular articles in the market. Simple bone dolls with movable limbs (No. 45), stuffed dolls, and jointed wooden dolls (No. 46) have come down to us. From a coffin of the late second century B.C. a jointed wooden doll was taken, complete with golden bracelets and a gold ring. The sarcophagus of the Empress Maria, daughter of Stilicho and wife of the Emperor Honorius (A.D. 384–423), contained a splendidly decorated little ivory doll. There were also larger carved wooden dolls meant to be dressed.

Authors of antiquity speak of automatons in human shape. About 350 B.C. Aristotle mentions in *De anima* and in *Politica* an 'automatic Venus'. Ptolemaeus Philadelphus is said to have shown an automaton at a bacchanalia in about 280 B.C. Hero of Alexandria (A.D. 110?) gained fame with his work *De automatis* in which he described turning and moving machines. Petronius writes in his *Symposium* of a silver doll which could move like a human being.

Though dolls' houses are mentioned neither by Greek nor by Roman writers, finds of very precious toy furniture made of bronze or ivory and many small vessels may well have belonged to dolls' rooms or houses. They are of the exact size usually found in dolls' houses—that is, about 1:10 to ordinary furniture. Karl Gröber takes them as sufficient evidence for the existence of dolls' rooms in Roman times. Pausanias, ancestor of all writers of travel guides, has described the beauty of the toy furniture of his day. He makes particular mention of a small doll's bed made of ivory which he saw in the Heraion at Olympia, and which Hippo-

43

44

45 *Dolls. Bone, movable limbs. Roman grave toy from Aquileia*

46 *Doll. Wood, movable limbs, gold ornaments. Roman, 2nd century* B.C.

47 *Dolls. Clay, movable limbs. Greece, 3rd–2nd century* B.C.

damia, wife of Pelops, had at one time dedicated to the goddess.

Few toys from the early days of feudalism in Europe have survived. Education in the Middle Ages was stern and unrelenting, and the world of the child of that time remains shadowy. The unrest of the period of the migration of nations also was not favourable to the production of toys. (An inventory of superstitions from the eighth or ninth century lists scrap dolls as *simulacra de pannis*.) The German wooden doll from the island of Usedom (about A.D. 1000) has been mentioned before (No. 13).

If during antiquity the social level of the doll was shown in the materials used, more or less precious, in the Middle Ages even more distinct types became characteristic of certain classes and social stations. Simple clay toys were sold by craftsmen at fairs to villagers and the working population of towns, while princely toys, like tilting knights and 'knightly dolls', were reserved for the children of the aristocracy and later on also the patricians. Toys for these classes corresponded entirely to the ideals of education of that stratum of society. According to their function in this society, and the need for a training to fulfil those functions, there were two main types of feudalistic education—one for the knights and the clergy, and another for the broad masses of the people. The training of boys is reflected clearly in the

45

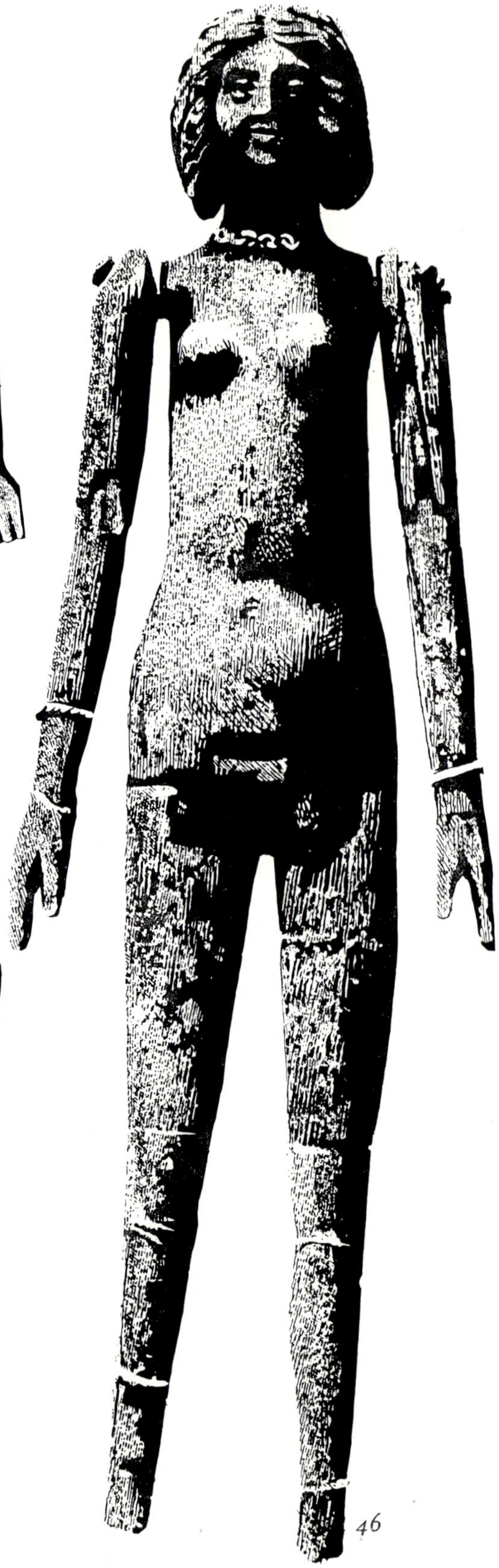
46

well-known illustration from the *Hortus deliciarum* by Herrad von Landsperg (about A.D. 1200). A picture shows two boys guiding on strings two knights fighting on foot, dealing each other blows with their swords. The figures are made of wood, jointed, and covered with cloth, a teaching model for future squires. Girls, too, were given dolls mounted on horseback, as is reported of Madeleine, daughter of King Charles VII of France. Even simple clay toys showed the influence of chivalry. Among the small dolls, clay horses, and knights in armour from the twelfth and thirteenth centuries found in Strasbourg, there was a mounted lady, carrying a hunting hawk on her hand (about 14 cm high). In Nuremberg a collection of toys was discovered under the pavement in 1859, including clay dolls up to 50 cm in height, which Karl Gröber believed to belong to the mid-fifteenth century (No. 50). Mass-produced, they were made of white pipe clay pressed into moulds and then fired. Their style is Gothic, and they usually have a little hollow on their chest which leads to the assumption that these dolls were perhaps christening presents, and that the hollow held the 'christening piece', a silver coin given to the child. The dolls show the costume of women of the period, long highwaisted pleated gowns, pleated sur-coats, and fashionable hairstyles.

As early as 1413 and 1465 the *Dockenmacher*—dollmakers—Ott and Mess are mentioned in Nuremberg records. The term *Dockenmacher* was at that time used for makers of toys in general, yet it seems certain that these two made mainly dolls. Nuremberg's reputation as the town of toys is founded on the skill of its craftsmen, and even more on the flair of its merchants. In the fifteenth century, favoured by its geographical position close to the main medieval trade routes, it was known as an important centre of international commerce. Its closeness to busy home industries, mainly in wood (Oberammergau, Berchtesgaden, Gröden), easy communications with the Sonneberg area, and a network of agents assuring distribution made Nuremberg an entrepôt for toys of all kinds.

The making of toys was a free craft not limited by guild rules. Highly skilled craftsmen made toys as a minor industry, with more than a dozen trades engaged: locksmiths, goldsmiths, wood-carvers, turners, metalworkers, joiners, and workers in pewter and brass. Later the famous dolls' houses of Nuremberg, Augsburg, and Ulm were made by the combined efforts of these craftsmen. Each of them was allowed only to handle the material appropriate to his trade: the joiner making furniture, the potter crockery, the goldsmith the jewellery of the small dolls. Christoff Weigel has often been quoted, describing the way the craftsmen worked, in his *Abbildung der Gemein-Nützlichen Hauptständ*, published in 1698:

48 Dolls and animal. Clay, incised ornamentation. Peru, c. 1930

49 Dolls. Ozokerite (mineral paraffin wax), cotton clothing, Sahara region, North Africa

> Dolls' houses contain everything necessary for a household's pride and ornament, daintily made, and sometimes precious... The material these toys are made of may be silver used by gold- and silversmiths, wood used by carvers and turners, or alabaster... Wax, too, is used, especially for the making of all kinds of animals and birds, made to look very natural, given a rough yet finely modelled skin daintily dressed with feathers—in fact there is hardly a trade in which the things usually made big are not often copied on a small scale, as toys.

Christoff Weigel also describes the important work of the maker of tragacanth dolls, who supplied a big popular demand. Tragacanth is a substance composed of vegetable gum, flour, and sugar, and these dolls were not true toys but edible like gingerbread figures and other sweetmeats. There was, too, the doll made of malleable material, produced by the man who also made masks from paper pulp, and painted them. The *Hortus sanitatis* (A.D. 1484) recorded the maker of wooden dolls with movable limbs.

The mass production of wooden dolls was decisively influenced by the introduction to toy-making of the turner's art. Christoff Weigel praised the turner, saying that he knew how to make many attractive dolls and toys. The turner was held in great regard everywhere, and his versatility and skill gave him a special place among craftsmen. During the sixteenth century, possibly even earlier, Nuremberg masters made rattle dolls on the lathe. These became prototypes for much fine folk art in the way of doll-making in many countries.

In the nineteenth century simply turned 'Dutch' dolls with movable limbs were exported from the Continent to Great Britain. Characteristic of German production are the turned baby dolls in swaddling clothes from Oberammergau, the rattle dolls and babies from Berchtesgaden turners, the wooden dolls and rattles from Thuringia (documented about 1700) as well as from the Erzgebirge (Nos. 58, 59). The economic manner of producing these toys allowed for a low price and wide distribution. Seiffen folk art has to a large extent developed from the basic form of the turned figurine. This can be seen in the candle-bearing miners and angels, the nutcrackers and pastil-burners. Simple turned toys, limited to certain typical characteristics, are particularly suitable for young children, since they stimulate the imagination. At the beginning of our century Seiffen toy-makers managed to alleviate the stiffness of their small dolls by using additional material (cloth, paper, feathers, hide) and extra colouring. Then the very stylized figures of different sizes became life-like. The miniature sizes (2–3 cm) were at one time among the cheapest toys. At present figure-turning is flourishing

50 *Dolls. Clay. Nuremberg, 15th century*

51 *Dolls*, kokeshi. *Wood, turned and painted. Japan*
52 *Dolls, group. Papier-mâché. India*
53 *Dolls, group. Coloured silk-worm cocoon. Korea*
54 *Dolls, group. Clay glazed. India*
55 *Funerary figure, life-size, representing the three-year-old Robert, Marquess of Normandy, died 1715. English (?)*
56 *Dolls. Wood (triangular in section), painted. Sagorsk, Russia, mid-19th century*

52

53

54

57 *Dolls. Wood, painted. Indonesia*

58 *Dolls. Wood, turned and painted, some with movable limbs. Erzgebirge, 19th century*

58

59 *Dolls. Wood, turned and painted. Erzgebirge, 19th century*
60 *'Dutch' dolls (figure on right 'Bohemian'). Wood, jointed. Val di Gardena. Italian Tyrol, 19th–20th century*

59

at Seiffen, and the making of well-turned figures for toys and fine wooden dolls ranks high in the training of workers in wood in the western Erzgebirge.

The turning of small dolls is not, as is sometimes supposed, a German specialty, but is found in various other countries all over the world, with national and regional variations in shape and ornamentation. It requires a mastery of turning techniques. There are the ninepin shaped dolls of Sweden, made of birch wood in the eighteenth century; the many turned babies in swaddling clothes with painted cradles, traditional in Czechoslovakia, with workshops in Dedová, Krouná and Skasov; the wooden figures of Russian folk art (No. 56); and the small modern Indonesian (No. 57) and Japanese dolls which are part of the general souvenir industry (No. 51). Almost a symbol of Russian folk art and an interesting souvenir are the *matreshka* dolls (No. 223). These cheerful, plump and gaily painted dolls, turned and hollow inside, are put one into another. There are sets of different sizes (4, 6, 8, 10, or 12 pieces).

The first Sonneberg dolls of the seventeenth century were turned and painted wooden dolls with carved wooden arms attached. Sonneberg turners, different from their colleagues in the Erzgebirge and in the Alpine countries, gave these dolls a dainty waist and narrow hips. Little glued-on noses made their faces look pointed.

61 *Dolls. Jointed, papier-mâché and linen. Japan*

62 *Bisque-headed doll. Kid body, movable limbs, wooden arms, mohair wig, sleeping eyes. Germany, c. 1870*

63 *Wax-headed doll. Leather body, wax limbs, natural hair. Britain, c. 1860*

64 *Bisque-headed doll. Wooden body, varnished, movable limbs, mohair wig, sleeping eyes. Germany, early 20th century*

65 *Bisque-headed doll. Kid body, wax limbs. France, c. 1870*

66 *Wax-headed doll. Kid body, wax limbs. Germany, 19th century*

Rattle dolls, too, were produced, their hollow bodies filled with dried peas or pebbles. Gayest among wooden figures were the dancing dolls. They stood on bristles, and were turned by the vibrations when put on a zither which was being played. The so-called *Ziehdocken* carry a baby in their arms, and by pulling a string threaded through the hollow body, the doll can be made to lift and lower her arms and with them the baby. There are turned dolls representing figures from daily life—for example, peasant women and dairy maids at the churn, these also with movable arms. At the beginning of the nineteenth century the shrewd Sonneberg doll-makers developed a wooden jointed doll. Its limbs could be moved, though the doll looked a little stiff and unnaturally slender. The wooden heads of these dolls have an imposed moulding of plastic material, and are given fashionable hairstyles. They are dressed in the style of the Empire period.

The Gröden doll, too, had wooden joints (No. 60). The great economic expansion of the toy industry in the Gröden valley during the nineteenth century was a result of the introduction of the mechanical lathe, extensive specialization within the industry, and the training of young people in wood-working. Wooden dolls took first place in the rich selection of work from carvers and turners which as early as 1810 was distributed by some 350 agents, big and small, in almost every country of Europe. Lithographed and hand-coloured catalogues and dealers' sample plates from the nineteenth century draw attention to, among other things, leather dolls, wooden jointed dolls, babies in swaddling clothes, and dolls' heads made of papier-mâché. It is on record that a carver produced one hundred dozen dolls in twenty-four hours, this mass production becoming possible only through the mechanical lathe. During the last eighty years native toy production has been largely supplanted by catering for an ever-rising tourist industry. Modern fraising machines now increasingly replace the carving tool, and make the term 'wood carving' very questionable. The Gröden wooden doll has become part of history.

The old German words for doll are *Docke* or *Tocke*. *Puppe*, a word borrowed from French, came to Germany probably during the sixteenth century via Alsace. In the seventeenth and eighteenth centuries it had almost ousted the older terms, which remained current only in the south of Germany. The minnesinger Oswald von Wolkenstein (1367–1445) called his beloved one his beautiful *Tocke*, or the *Tocke* of all his joys. Neidhardt von Reuenthal (about 1180–1250) called his Vriderun, a *Tocke*, and even Martin Luther (1483–1546) spoke, regarding feminine vanity, of women as 'pretty *Tocken*'. But Johann Fischart (1547–1589 or 1590) and Geiler von Kaiserberg (1445–1510), both from Alsace,

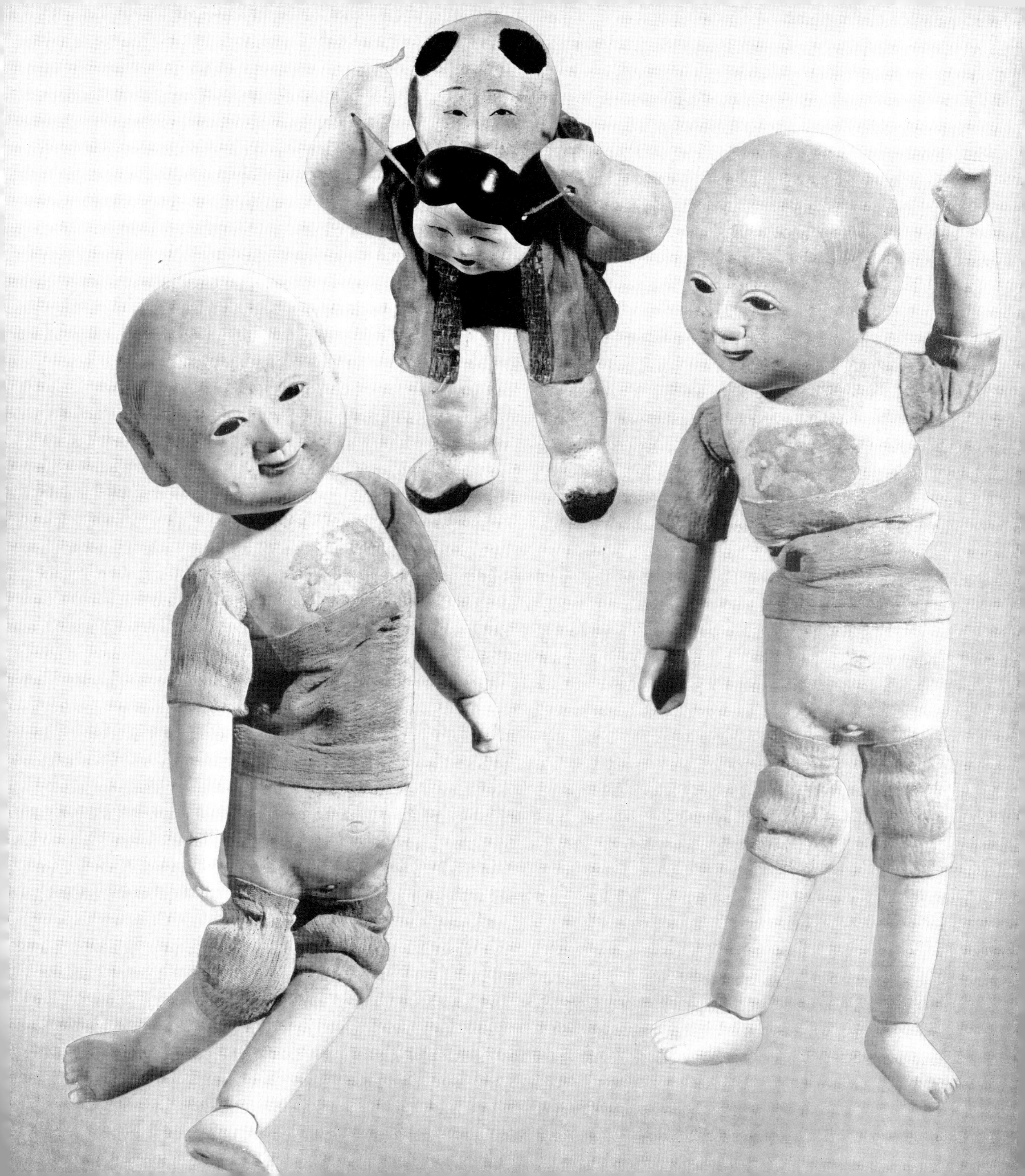

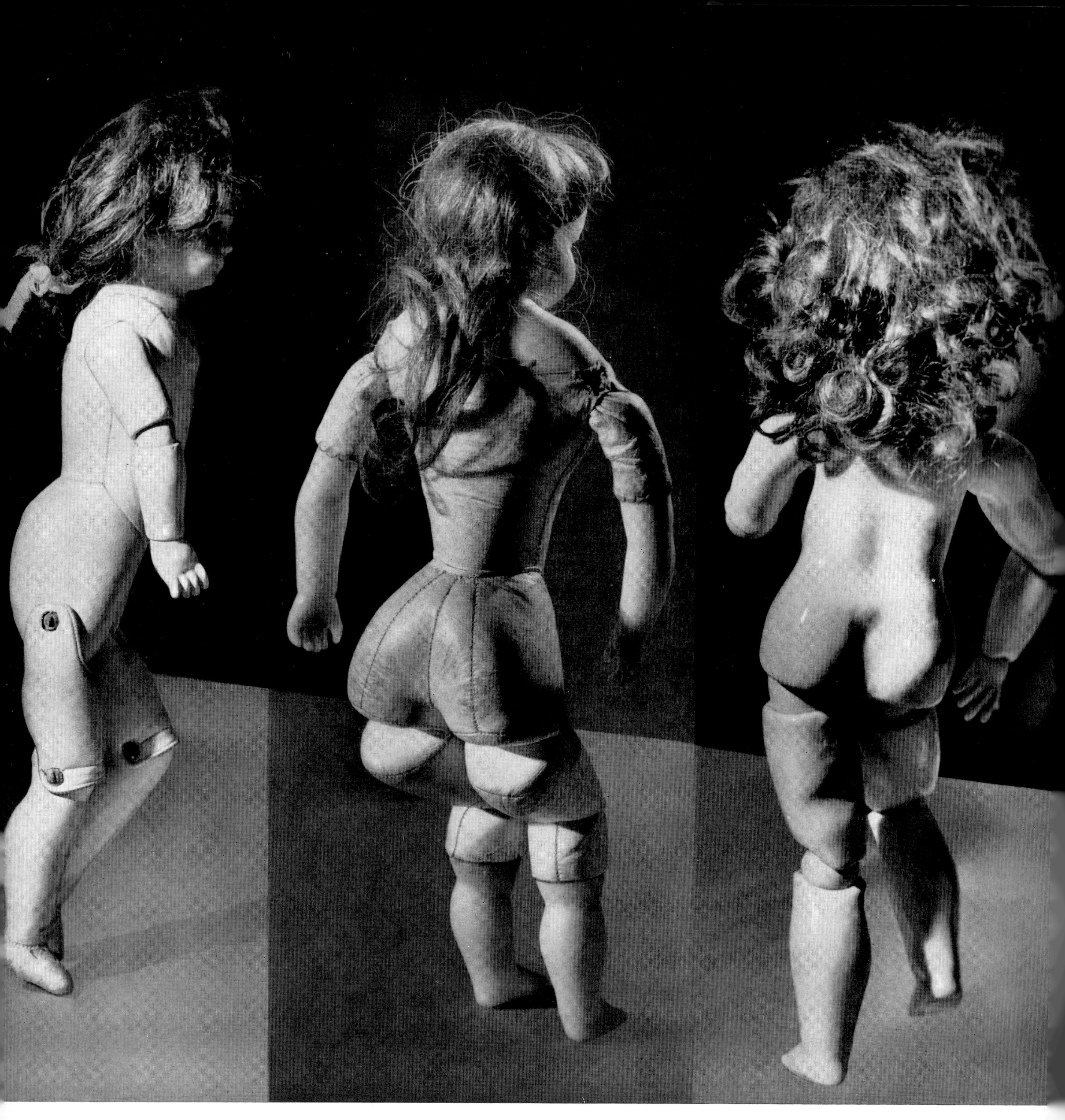

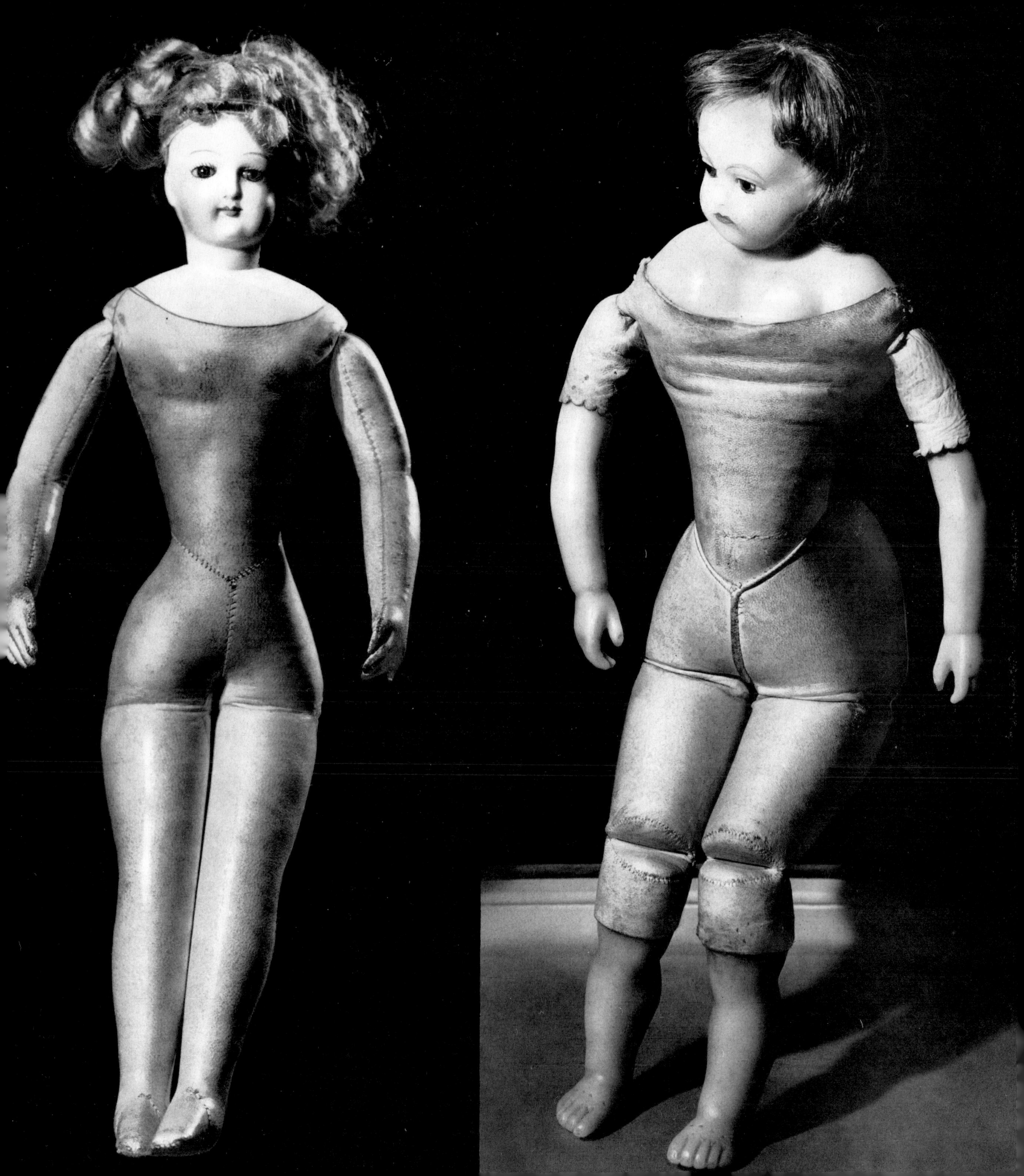

67 Life-size figurines by Sasha Morgenthaler. Moulded material under stockinette, glass eyes, hair made of cord. Zürich, 20th century

68 Dolls. Birchbark, Sweden, 19th century

69 Votive doll. Dried palm leaves. Used as a toy after losing religious significance. Bali, late 19th century

68

preferred in their writings the word *Puppe*. The *Frauenzimmerlexikon* of 1715 uses both terms in a section on toys.

In many countries dolls of easily available materials belong to peasant tradition and folk art, made often by the country people during their enforced leisure in winter to earn some extra money. To this group belong the Russian moss-men (No. 233); roughly carved, sometimes painted figures of Alpine herdsmen in Switzerland; and figurines made of fruit for special occasions, like the *Pflaumentoffel*, the little prune-men of Saxony. The attractive small birchbark dolls belong to Swedish folk art, and were produced as supplementary sources of income by many villagers (No. 68). Birchbark could be pleated or woven, the technique itself being ornamental. The surface of this pliable material was covered with attractive designs. It was used mainly in those parts of Sweden cultivated as long as four hundred years back by the Finns. The material displays its decorative qualities well as a base for small dolls in peasant costume, which bear witness to the skill and sense of beauty of simple people.

3 *The Doll as a Mirror of Social Realities*

A history of the doll highlights at the same time the history of fashion and the changing idea of beauty. It gives a clear picture of social developments, and above all displays the ideal of feminine culture of any epoch. Until the turn of the eighteenth to the nineteenth century the child was regarded as a small adult, unfinished and hence to be trained in manners. Children's clothes were small, stiff, and slavish copies of adult dress. This was particularly true of the children of the wealthy. The children of the poor, as a rule, knew only four pieces of clothing: a smock, a skirt for the girls, and trousers and a jacket for the bigger boys. These completely lacked splendour and ornament. The young of the upper classes, the future lords and ladies, were, from the fifteenth century onwards, dressed like their elders, wearing in turn heavy dresses, elegant lace collars and frills, wasp waists and farthingales or tightly fitted clothing which impaired breathing and movement. Bonnets, hats, hairstyles, wigs, and shoes aped the dress of their 'betters' whom the children were expected to follow in almost very respect.

The many portraits that have come down to us from the past show clearly how children reflected the society in which they lived, its faults and its virtues. This is particularly obvious in the formal portraits fashionable from the sixteenth to the eighteenth century. To the modern mind these posed and lifeless attitudes are the very negation of everything children are. But that was their world: in habits, clothes, speech and ways of thought children were miniature unfinished adults, and recognized as such. The writer Johanna Schopenhauer, mother of the German philosopher, has described in her memories of childhood the affliction of being fitted into one's clothes as into heavy armour. She refers to the last thirty years or so of the eighteenth century, saying that an enormous tower of hair added over a yard to one's height. This was based on a wire frame and horsehair, crowned by enormous masses of feathers, flowers, and ribbons. Stilt-like heels on bow-adorned dancing shoes further added to the discomfort, as only the tips of the toes touched the ground. Then there was a corsage made of closely fitted whalebone, firm and stiff enough to repel a musket ball. This pressed forcefully on arms and shoulders, pushing them back, and the chest forward, at the same time reducing the waist above the hips to wasp shape. And then there was also a hooped skirt!

As long as clothing kept to the reasonably natural standards set by antiquity, as it did about up to the Middle Ages, the need for special dress for children did not arise. Only when children's dress became less and less comfortable and more and more complicated—one thinks of the fashions of Burgundy, Spain, and the

70 *Bread dolls. Valais, Switzerland, 19th century*
71 *Doll. Wood, painted, natural hair, painted eyes. Britain, 19th century*
72 *Soft doll. India, 20th century*
73 *Soft doll. Painted face, natural hair. Sardinia, early 20th century*
74 *Dolls. Wood, national dress. Morocco, present day*
75 *Doll representing Red Indian. Head made from a dried apple. Present day*
76 *Doll representing Appenzell peasant. Wood. Switzerland, late 19th century*
77 *Doll representing Eskimo woman. Wood and sealskin. Baffin Island, early 20th century*

70

Rococo period—did a demand for easing the strict styles become vocal. Clothes for children as a separate category are known only from the second half of the eighteenth century in Britain, influenced, it would appear, by the revolutionary theories of education put forward by Jean-Jacques Rousseau. He considered play a better training for the young than their constant consorting with adults. In 1762 Rousseau warned in his novel *Emile* of the danger of destroying the child's world. He called for simplicity, including simple things for the child to play with. Small twigs with their berries and leaves, he thought, a poppy head with its seeds rattling, or the sheer comfort of a stick of liquorice could on many an occasion far outweigh the most expensive bauble ever produced.

The age of the French Revolution brought about a change in the dress of adults, also. A dispensation by the Emperor Joseph II, dated 1783, prohibits girls in orphanages, convents, and other places of education from wearing tightly fitting bodices. Contemporary almanacs make suggestions for sensible children's wear—for example, the collar *à la matelot* which was to free boys' necks from the confines of the ruffle and the jabot. Also recommended were easy shift dresses *à l'anglaise*, and soft, flat shoes for girls. But with the coming of the *Biedermeier* period in Germany these good beginnings were forgotten in a new wave of styling

78 Doll's head. Papier-mâché, stucco finish, silk hair. Japan. First quarter of the 20th century

79 Doll's head with historical hairstyle. Japan, 20th century

80 Baby doll in christening robe. Wax sleeping eyes, natural hair. Britain, c. 1880

80

79

81 *Girl dolls. China, early 20th century*

82 *Wax-headed doll. Leather body, bust and hands wax, wig modern. France, c. 1784*

83 *Doll. Head papier-mâché. Germany, first half of the 19th century*

children's clothes on those of adults. About 1850 the fashion of the hooped skirt, the so-called crinoline, again forced girls and their mothers into whalebone structures, though the skirt had sensibly shortened. Boys' fashions of that time show military influence. Though demands of reformers were slowly gaining ground in the nineteenth century, they were not fully heeded until the twentieth, the much-lauded century of the child.

The paternal system of education which lasted well into the nineteenth century was the reason for the doll—excepting the simple wooden, carved, turned, or soft baby doll—being patterned on the adult. Rules for play were often didactic and to our ideas devoid of understanding of the child's own world. Even the doll was a model to instil respect, at best a means to training the future wife and mother. To this day in Japan traditional dolls teach the child the social order of rank and the history of the country. Among the *hina ningyō* which are put up at the famous Doll Festival the emperor and empress take first place (No. 97).

About 1809 Friedrich Campe of Nuremberg published a book, *Kinderfreuden*, which described itself as a pleasant picture book for the young. It is, in fact, a typical example of the then current pompous and windy writing for children, yet it is interesting for the light it throws on the manners and mores of the period. Describing one of the illustrations it says:

On this page we see a party of young girls playing with their favourite toy, a doll. They sit comfortably engaged in dressing their playthings. They sew for them, thus introducing themselves almost unnoticeably to their own future calling. Rough games do not suit little girls. But a quiet domestic sense, diligence and a wide knowledge of all the work required in a home, are the foundations upon which their chance of future happiness must be built. And what could better prepare them in this than doll play? The doll is the helpless little being they must care for. For it all kinds of dresses, hats and underwear have to be made. Dolls have to be dressed and undressed, even their small beds must be made. Dolls also have their kitchens, with every kind of cooking utensil . . .

All this leads young girls into their sphere in life, it early gives them a firm directive for their feminine qualities, and, if their parents are sensible and arrange their play carefully, these early lessons will survive through life.

So you, dear little girls, to whom this book is given, should always regard playing with dolls as a means to being good and happy in your future life.

Writing of this kind, divorced from the age in which it was written, seems amusing to a generation which no

84 Porcelain-headed doll. Historical costume. Germany, 19th century

85 Doll. Wood, painted. Britain, c. 1715

85

longer regards a woman as condemned for life to the wash tub. But at the time, and before and long after, it referred to a way of life which no one questioned: the woman depending on man and being the object of his exploitation and desire.

Until 1800 most dolls tended to represent the outlook of the wealthy upper-class customer. The splendid dolls' dresses, no less costly in material and ornament than the clothes on which they were modelled, made such toys suitable for princely presents, and outside the reach of the average purchaser. His children had to be contented with simple dolls of clay or wood, sold cheaply at fairs. In the sixteenth to the eighteenth century the dolls' house was a pastime of the wealthy only. In the nineteenth century the miniature began to assume its present function as a true dolls' house being based on middle-class domesticity.

It was left to the mid-nineteenth century to popularize the child itself with the baby doll.

The great number of precious dolls in museums and private collections might easily obscure the fact that the average child played happily with much simpler toys. Hubert Kant points this out in the introduction to the catalogue for the exhibition *Das Kind und seine Welt* in Vienna in 1960:

> The world of the child . . . is above all the world of the middle-class child for it is from this sphere that most evidence has survived. This shows up the bright side of the century of the middle-class. . . . The dark side of the world of the working class child was scarcely touched by the brilliant play culture of the eighteenth and nineteenth centuries. Through lack of evidence and pictorial representation it can only be guessed at yet has to be taken into account.

Class contrasts in society must always be reflected in the world of the doll, and there is a vast difference between the doll with trousseau given by the Duchess

86 *Character dolls. Modelled heads, painted, cloth bodies, movable limbs. England, c. 1900*

87 *Dolls. Plastic. German Democratic Republic, present day*

88 *Wax portrait doll. Queen Victoria in her coronation robes. Britain, c. 1858*

89 *Figurine. Wooden head, tow hair, tow-stuffed body. It is doubtful whether this figurine, representing a soldier, is a doll. 18th century*

of Orleans in 1722 to a French princess at a cost of about 20,000 francs, and those simple dolls based on folk tradition showing the hard workaday life of the people whatever ethnic regions they may belong to. There is plenty of evidence however in life and literature that the preciousness of a toy does not necessarily add to its play value. Charles Baudelaire (1821–67) and Jean Paul (1763–1825) give examples in their works, and a most perceptive description of children at play comes from the Swiss writer Gottfried Keller (1819–90) in his charming tale, *A Village Romeo and Juliet*. The two are crofter's children, and their doll is made of the simplest of materials:

> The two children, on the other hand, who had already made plans to return home with their fathers, drew their cart under the shelter of the young lindens, and betook themselves on an excursion into the waste field, which to them, with its weeds, bushes and heaps of stones, represented an unfamiliar and fascinating wilderness. After they had wandered hand in hand for a little while in the midst of this green waste, and had amused themselves by swinging their joined hands above the high thistles, they finally sat down in the shade of one of these, and the little girl began to make a dress for the doll out of long plantain-leaves, so that it was presently decked out in a beautiful, green, scalloped skirt. A solitary red poppy, which was still blooming there, was drawn over its head as a hood and fastened on with a blade of grass; all of which made the little figure look like a witch, especially after it had acquired, in addition, a necklace and belt of little red berries. Then it was placed high up on the stalk of a thistle, and contemplated for a while with united glances, until the boy had looked at it long enough and brought it down with a stone. This completely disarranged the doll's toilet, and the girl speedily disrobed it in order to dress it anew. But the moment the doll was again quite naked and rejoiced only in the red hood, the wild youngster snatched the plaything away from his companion and threw it high into the air. The girl ran after it with a cry, but the boy got it first and gave it another fling. He continued to tease her in this way for a long time, the little girl all the while struggling vainly to get hold of the flying doll, which suffered considerable damage at his hands, especially to the knee of its one leg, where a small hole appeared, allowing some bran to trickle out. As soon as the tormentor noticed this hole, he stopped still as a mouse and with open mouth eagerly began to enlarge the hole with his finger-nails, in order to ascertain the source of the bran. His standing still seemed highly suspicious to the poor little girl; she crowded up close to him and with horror was compelled to witness his wicked proceeding.

90 *Dolls representing Lapp couple. Early 20th century*
91 *Doll representing Eskimo in kayak. Sealskin, with complete hunting tackle. Greenland, late 19th century*

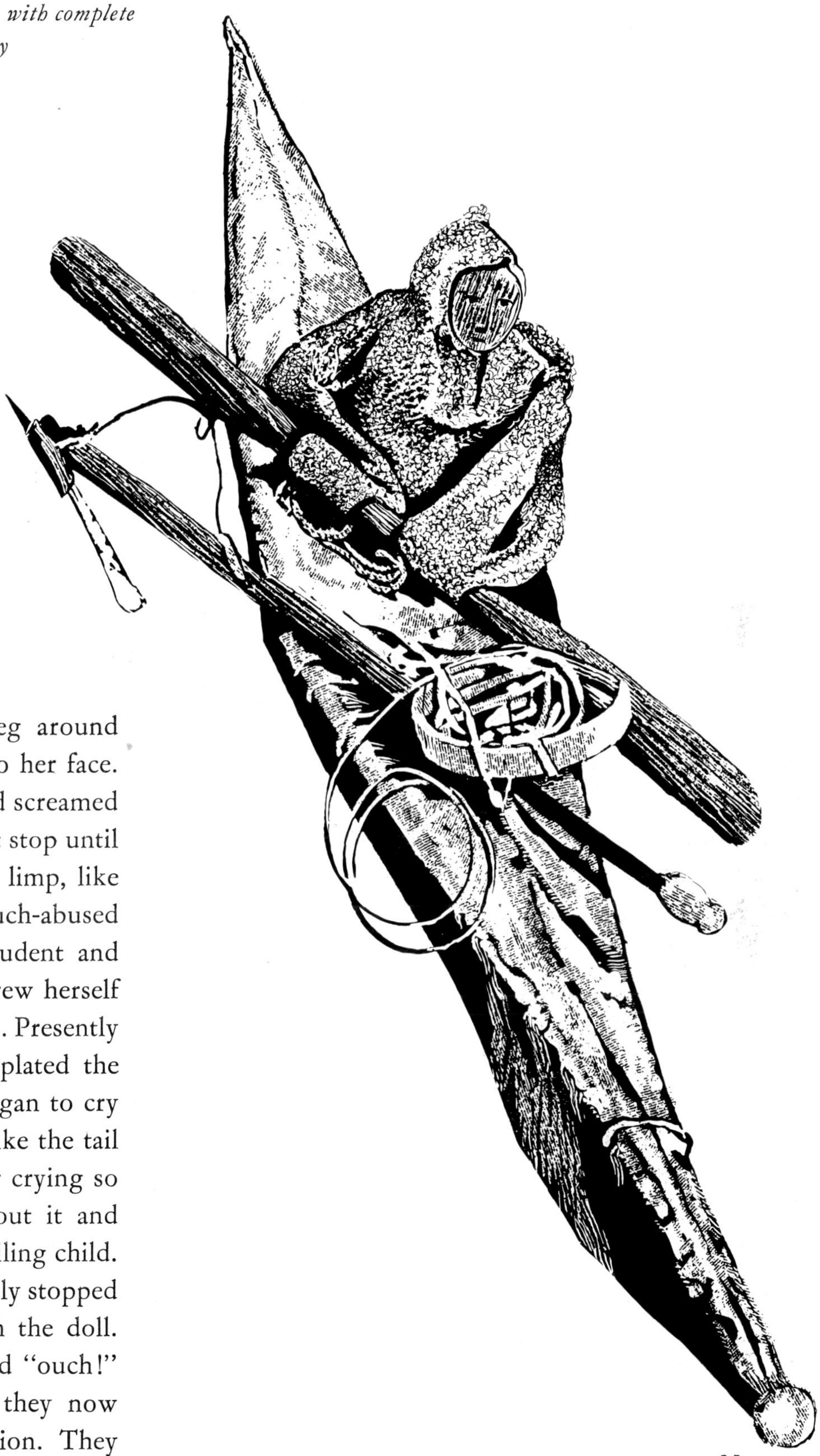

91

"Oh, look," he cried, and jerked the leg around before her nose, so that the bran flew into her face. And when she made a grab for the doll and screamed and begged, he ran away again and did not stop until the whole leg was empty and hung down limp, like a pitiful husk. Then he flung down the much-abused toy once more and assumed a very impudent and indifferent air, as the little girl, crying, threw herself upon the doll and covered it with her apron. Presently she drew it out again and sadly contemplated the poor thing. When she saw the leg she began to cry afresh, for it hung down from the body like the tail of a salamander. The evil-doer, seeing her crying so violently, finally began to feel badly about it and stood anxious and repentant before the wailing child. When she became aware of this, she suddenly stopped crying and struck him several times with the doll. He made believe that it hurt him and cried "ouch!" so naturally that she was satisfied, and they now resumed together the destructive dissection. They

92 *Astronaut dolls. Plastic. Moscow Toy Factory, present day*

93 *Figure of knight. Figure and horse's body wood, armour metal. Southern Germany, c. 1530. The crest is that of the patrician family Holzschuher, Nuremberg*

92

bored hole after hole into the little martyr's body, letting the bran run out from all sides. Then they carefully gathered it into a little pile on a flat stone, stirred it around, and looked at it intently.

The one intact member which the doll still retained was its head, and of course this now attracted their chief attention. They severed it carefully from the eviscerated body and peeped with amazement into the hollow interior. When they saw the ominous hole and also the bran, their next most natural idea was to fill the head with the bran. And so the children's small fingers began to vie with one another in putting the bran into the head, which now, for the first time in its existence, had something in it. The boy, however, probably still regarded it as dead knowledge, for he suddenly caught a large blue fly, and holding it, buzzing, in the hollow of his hands, he ordered the girl to empty the head of bran. The fly was imprisoned in it and the hole stopped up with grass, and after both the children had held the head to their ears, they solemnly put it down on a stone. As it still had on the red-poppy hood, the resonant head now resembled a prophet's poll, and the children with their arms around each other listened in profound silence to its oracles and fairy-tales.

But prophets always awaken terror and ingratitude; the bit of life in the poorly formed image aroused the children's human propensity to be cruel, and it was decided to bury the head. Without asking the imprisoned fly's opinion about it, they dug a grave, laid the head in it, and erected an imposing monument of field-stones over the spot. Then a gruesome feeling came over them, as they had buried something with life and form, and they went some distance away from the uncanny place.

This incident is related of Switzerland, but, as we have seen, a liking for simple playthings is universal; a child in a remote Tyrolean village was discovered playing with a doll consisting of an electric-light bulb

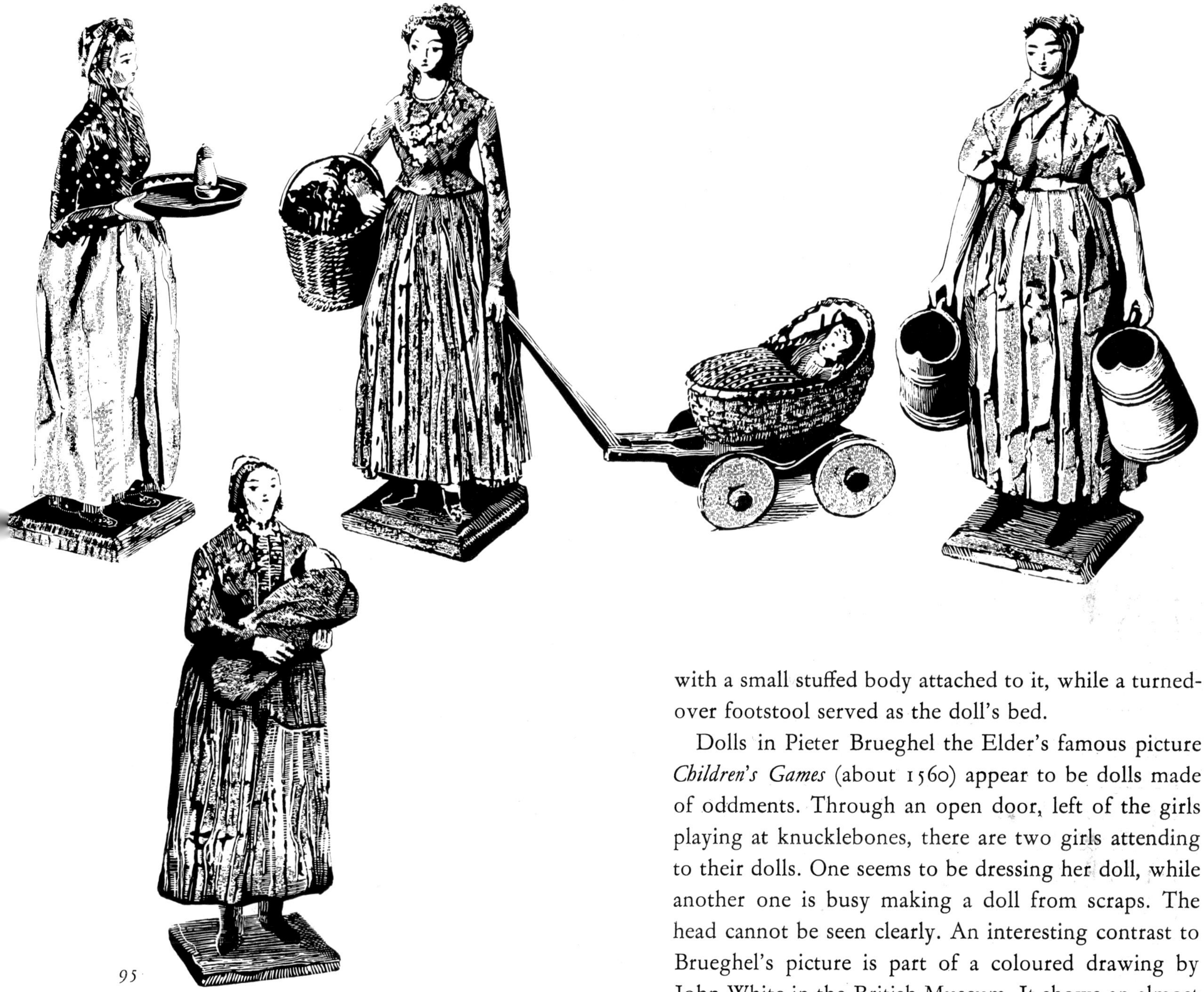

95

with a small stuffed body attached to it, while a turned-over footstool served as the doll's bed.

Dolls in Pieter Brueghel the Elder's famous picture *Children's Games* (about 1560) appear to be dolls made of oddments. Through an open door, left of the girls playing at knucklebones, there are two girls attending to their dolls. One seems to be dressing her doll, while another one is busy making a doll from scraps. The head cannot be seen clearly. An interesting contrast to Brueghel's picture is part of a coloured drawing by John White in the British Museum. It shows an almost naked Red Indian child standing beside her mother. In her hands she holds a doll dressed in Elizabethan fashion. In 1585 English colonists landed on Roanoke Island along the coast of North Carolina. Among the presents and goods for barter with the Indians were dolls. White's drawing is one of the earliest pictorial representations of European toys in the New World.

Charles Dickens (1812–70) vividly describes a toy-maker's workshop in *The Cricket on the Hearth*:

94 *Religious dolls, dressed as nuns. Italy, 19th century*
95 *Paper dolls. Dresden, c. 1830*

96 *Dolls,* Staatsdamen. *Sonneberg, 19th–20th century*
97 *Tableau of the Japanese Doll Festival, 20th century*
98 *Doll dressed as bride of the civil-servant class. Clay, soft body, gold-embroidered dress, bead-decorated headdress. China, 20th century*
99 *Dolls, group. Large doll, bisque head, glass eyes. Small doll, bisque head, painted eyes. Sonneberg, 19th century*
100 *Cook in the kitchen. Originally a crib figure. Germany, 2nd half of the 19th century*

Caleb and his daughter were at work together in their usual working-room, which served them for their ordinary living room as well; and a strange place it was. There were houses in it, finished and unfinished, for Dolls of all stations in life. Sub-urban tenements for Dolls of moderate means; kitchens and single apartments for Dolls of the lower classes; capital town residences for Dolls of high estate. Some of these establishments were already furnished according to estimate, with a view to the convenience of Dolls of limited income; others could be fitted on the most expensive scale, at a moment's notice, from whole shelves of chairs and tables, sofas, bedsteads, and upholstery. The nobility and gentry and public in general, for whose accommodation these tenements were designed, lay here and there, in baskets, staring straight up at the ceiling. But in denoting their degrees in society, and confining them to their respective stations (which experience shows to be lamentably difficult in real life), the makers of these Dolls had far improved on Nature, who is often forward and perverse; for they, not resting on such arbitrary marks as satin, cotton-print, and bits of rag, had super-added striking personal differences which allowed of no mistake. Thus, the Doll-lady of Distinction had wax limbs of perfect symmetry; but only she and her compeers—the next grade in the social scale being made of leather, and the next of coarse linen stuff. As to the common people, they had just so many matches out of tinder-boxes for their arms and legs; and there they were—established in their sphere at once, beyond the possibility of getting out of it.

All schools of thought have for long agreed on the importance of play, maintaining that play is education. The doll in particular helps to form basic attitudes and behaviour in the child, towards the people of his immediate surroundings, family first, and then society in general.

The idealistic biological school sees in play an emergence from instinct, the materialist a compulsive attempt of the child to bring to his level the world of adults around him, which is in short a deliberate, if unreasoning act.

In the sixteenth century Erasmus (1466–1536), the Dutch scholar, held that while it was never too late to learn, a child's impulse to imitate allowed it to learn instinctively what could only be acquired later at greater cost and exertion. In the nineteenth century Froebel (1782–1852), German educator and founder of the kindergarten system, strongly advanced the theory that play was in fact an early form of mental discipline, and that through it the world could alike be discovered and conquered.

101 *Pedlar doll. Wooden body, wax head, natural hair, painted eyes. Britain, between 1835 and 1849. Stall has been much added to; probably not originally with doll*

102 *Pedlar dolls. Wooden bodies and heads. Legless doll, wax head. Britain, first half of the 19th century*

102

103 *Baby doll. Wooden head, cradle leather-covered wood. Lapland*
104 *Wax-headed doll. Cloth bust and wax arms only, natural hair. Home-made, with silk dress. 20th century*
105 *Figurine. Wood, painted, tow hair. Bavaria, 18th century. It is doubtful whether this figurine, representing a courtier, is a doll*

Makarenko (1888–1939), the classical representative of Marxist education, stressed the same theory even more strongly. He maintains that play in the child has the same significance as has occupation, work and service to the community in the life of the adult.

Codified, the whole viewpoint was expressed by Maxim Gorki (1868–1936) when he said: "Play is the child's way of getting to know the world in which he lives and which he is called upon to change."

The Middle Ages produced those fine dolls for the training of boys, as shown by two tilting knights (about 1530) with precious helmets and beautifully made armour (No. 93). There is also an eighteenth-century soldier with horse made of wood with a tow-stuffed body (No. 106).

In 1516 Hans Burgkmair made his well-known woodcut of the games of the later Emperor Maximilian. The Emperor remembered well the games of his childhood and did much in his day to revive the spirit of chivalry. The Augsburg armourer Koloman made two horsemen similar to those in Burgkmair's picture for Maximilian's ten-year-old son Ludwig.

The role of background in conditioning play is indicated in an account by the Russian educationist, Konstantin Dimitriyevitch Ushinski (1823 or 1824 to 1870):

With one girl the doll bakes, sews and irons, with another she sits on the sofa, receives visitors, hurries to the theatre or to a party. With another again the doll beats her servants, has a savings bank, and counts her money. We have met children whose gingerbread men occupy certain ranks and take bribes.

In the First World War soldier dolls and tin soldiers sometimes served in the chauvinistic and militant training of the young. The spirit of war entered the nursery. Käthe Kruse herself produced 6–8 cm high figures of soldiers, inspired by photographs from camps and battlefields. The figures, Germans and 'adversaries', were covered with cloth and were movable, and could be marshalled into any position desired. There were in Britain at the same time rag dolls of soldiers, some 35 cm high, their bodies consisting of bits of printed cotton, sewn together and then stuffed with various materials (No. 107).

In North America, Canada, and Greenland the Eskimos still make dolls of seal hide. Equipped with fishing tackle, they reflect the hard struggle for existence on Northern coasts (Nos. 77, 91). Since the beginning of Soviet space exploration in 1961 the Moscow toy factory has been producing gay astronaut dolls from plastic materials, which are exported to many countries,

107

106 Figurine representing a dragoon. Tow-stuffed body, head and limbs wood, woollen clothing, wooden horse. Bavaria, late 18th century

107 Cut-out dolls. Printed on linen, to be sewn and stuffed. Britain, 1915

including the U.S.A., Britain, France, India, and Japan (No. 92). These simple space travellers are pleasantly different from the astronaut 'supermen' with space guns and laser pistols which populate the toy shops of some countries (such as *Major Matt Mason*, Mattel's man in space).

In the eighteenth century dolls representing nuns and priests were made in Italy and Germany, among other places. In Catholic regions they were given to children, encouraging them to act out in play religious ceremonies (No. 120). The tin-smiths made small rosaries, amulets, monstrances, incense burners, candelabra, and sanctuary lamps. The wood-workers of Berchtesgaden made praying monks and nuns as figures to be opened. Inside they were painted with pictures of a figure kneeling at prayer (about 5 cm high). The German Romantic poet Clemens Brentano (1778–1842) has described how in the lumber-room of a house he came across all the paraphernalia of 'past Christmases', including some dainty crib figures made of wax; monks and nuns, church dignitaries, popes and hermits. Small toy altars made of wood, and frequently fitted with a musical box, were popular mainly in Latin countries until recently.

During the nineteenth century the toy industry made increasing use of paper and cardboard. The ten paper-dressed dolls in the Staatliche Museum für Volkskunst in Dresden (No. 95) belong to the finest achievements of their period. They have been mentioned repeatedly in literature on toys and shown in pictures. These little figures in the round are on average about 21 cm high, and have wooden bodies fixed to a wooden base by a peg. The faces and necks are made of pulp, probably cast in moulds, then fixed to the wood like a mask and painted. The feet, too, are made of pulp (probably bread dough), and patterned coloured paper, possibly cut from bits of wallpaper, is used for the dresses, which must have been put on while still wet. Though no records regarding these dolls have been found so far, the technique appears to show that the dolls were mass

108 *Doll. Painted wooden head, wooden hands and feet, leather-covered. Britain, between 1695 and 1710.*

produced. The museum's inventory unfortunately gives only slender information—namely, that in 1835 the pram in the little collection there was played with by "the daughter of the Court Watchmaker, burgher and householder Johann Andreas Ludwig Täubner, Rampische Strasse".

The dolls represent women in contemporary middle-class dress, busy with household chores or doing their shopping. Everyone of them carries something: pitchers, a handbag, a shopping bag, a wine basket, an ordinary basket, a tray, or a baby. Some appear to be servants. All details are lovingly executed, placing these dolls, as Karl Gröber has pointed out, in the same category as the paper-made religious articles, known to be the products of nunneries.

In earlier descriptions of the Dresden dolls none of the authors has apparently discovered that the tray-carrying figure is directly borrowed from fine art—namely, from the famous *Schokoladenmädchen*, a pastel drawing on parchment in the Dresden Gemäldegalerie. It was done at the Austrian Court soon after 1743 by Jean-Etienne Liotard (1702-89), and represents a maid gracefully carrying a tray with a cup of chocolate and a glass of water.

The pedlar dolls of the second half of the eighteenth and the first half of the nineteenth century are a typically English phenomenon. Our illustrations show the many existing types. They are usually made of wood, some with wax heads, and represent old women, or more rarely men or young girls in the working dress of the people, and are in strong contrast to fashionable dolls. In baskets hung around their necks, on trays, or on a stall, in imitation of real pedlars in the countryside, they offer haberdashery, household goods, toys, and other useful things. All these articles were made to scale in miniature. The real pedlars also provided lonely housewives with news, gossip, mottoes, and fortune-telling slips. This led in the late eighteenth century to the appearance of the somewhat better-dressed fortune-telling doll. Her skirt was made of paper, and contained fortune-telling slips, written in ink. These popular dolls reflect contemporary customs, and often show a small label with a name and date on the tray, just as the real pedlars had to display their names and dates of licence in a visible place.

The toy seller among the Nodding Figures from the Rhön, with a movable jaw (early nineteenth century), and the Slovak Tinker, a popular pastil-burner from the Erzgebirge, are German parallels to the pedlar dolls, conveying a glimpse into social history, as do the Street Criers of London and Paris, made of Dresden china during the Kändler period (1731–65). They include hawkers, street-sellers, jugglers, buffoons, beggars, wandering preachers, and quacks.

4 *The Doll Culture of the Eighteenth Century*

Ever since the Middle Ages dolls were sent from France to the European Courts as welcome ambassadors of fashion. They publicized the latest French styles, true in every detail. In the nineteenth century fashion-plates and journals took over this function much more cheaply. Fashion dolls appeared for the first time in the accounts of the French Court when the Court Embroiderer, Robert de Varennes, in the service of Charles VI, received 459 francs for a doll's wardrobe he had made. Queen Isabeau de Bavière (1371–1435), born a Bavarian princess, sent this to her daughter, Queen Isabella of England. Before Henri IV of France's second marriage he sent some dolls to his bride, Maria de Medici, explaining: "Frontenac tells me that you wish to have some models of our fashions."

Certainly French fashion dolls, which sometimes reached life-size, and were more richly dressed than toy dolls of the same period, were due to the demands of the Courts. In the seventeenth century the obtaining of fashion news through dolls seems to have been fairly widespread. Johann Michael Moscherosch (1601–69) of Strasbourg in 1640 wrote a dream-satire, aimed at the 'establishment' accepted idea of baroque, *Wunderliche und wahrhaftige Gesichte Philanders von Sittenwald*. In it he made fun of German women who all too readily accepted Paris fashion. An anonymous lampoon of 1689 says, "And the worst is that women do not only travel to

109 *Doll. Moulded head with painted eyes. Paris, c. 1750*

110 *Automaton, mandolin player. Wax head, wooden hands. Italy (?), pre-1760*

111 *Venetian puppet, 60 cm high. Wooden head, hands and feet, velvet and silk damask clothing. Mid-18th century*

France but have models, dressed dolls, sent which cost many thalers, so that they can imitate carefully the devil's vanity." The fashion doll reached its greatest popularity in the eighteenth century. The ostentatious dissipated life at the numerous Courts with their extravagant feasts, and the increasing travel of the nobility and the wealthy middle class furthered the expensive custom. *La Poupée de la rue St Honoré*, centre of Paris fashion designers, became *Le Grand Courier de la Mode*. With the help of the mannequin export grew into a lucrative business for French fashion houses. Special shows with luxuriously dressed and coiffured dolls vied for the customers' favour. Among society ladies it became the done thing to keep a couple of elegantly dressed dolls as a pastime, one in formal attire, *grande toilette*, and the other in elegant *négligé*. These dolls were called the 'big Pandora' and the 'small Pandora'. Great Britain proved a particularly receptive market for the French fashion trade. In spite of the hostilities during the Spanish Wars of Succession (1701–13) both countries allowed the *mannequin* free entry as "an act of gallantry towards the ladies". All this is related in a special report by the Abbé Prévost from the year 1704. In 1712, that is to say, when the war was still on, English papers stated: "Last Saturday the French doll for the year 1712 arrived at my house in King Street, Covent Garden." Queen Marie Antoinette's famous milliner, Rose Bertin, made dolls according to the latest fashion for the Queen and her mother, the Empress Maria Theresa. Madame Bertin's dolls were coveted objects in all European capitals, from Paris to St Petersburg. After the French Revolution Rose Bertin opened a business in London for her client among the French emigré aristocracy. French and most likely also English fashion dolls soon crossed the Atlantic to America. The *New England Weekly*, of July 12th, 1733 carried this announcement:

> To be seen at Mrs. Hannah Teats, mantuamaker, at the head of Summer Street, Boston, a mannequin dressed after the newest fashions of mantuas and nightgowns and everything appertaining to women's attire, lately brought on the *Captain White*. Ladies who choose to see it, may come or send for it. It is always ready to serve you. If you come it will cost you two shillings, but if you send for it seven shillings.

The enormous social changes brought about by the French Revolution were clearly reflected in costume. Dress became both more comfortable and more functional, and at the same time fashion spread to wider circles of the population. The influence of Britain and the United States now made itself strongly felt. The *mode à l'anglaise* penetrated to France, and arrived as *Werther* costume in Germany. In France the dress of the third estate, the rich bourgeoisie, now determined developments, and fashion dolls geared to the demands of the Courts lost their meaning, though during the first

112 *Figurines. Wax heads, tow-stuffed bodies. Woman in the costume of Munich, c. 1800. Male figure c. 1760. Southern Germany, 18th–19th century*

113 *Doll. Wood, movable wooden legs, natural hair. Costume of 1755. Britain, 18th century*

114 *Group. Lady and manservant on a sleigh. Wooden heads, hands, and feet, tow-stuffed bodies, silk and brocade clothing. Southern Germany, 18th century*

115 *Figurines in fashionable dress. Germany, 18th century*

half of the nineteenth century they were still in use, adapted to current fashions.

A revolutionary invention conquering the market from 1790 onwards was the flat doll made of paper or cardboard, the paper doll for cutting out. Picture sheets with typical representations of different professions and trades and of soldiers were popular. In Germany such sheets in the form of coloured woodcuts had existed since around the middle of the seventeenth century; a varied wardrobe for the small dolls was also provided on the sheets. The English paper doll is sometimes said to have fulfilled a dual purpose—that of a children's toy and a fashion sheet. Between 1785 and 1827 there appeared in Weimar the first German fashion journal intended as a source of information for 'people of quality'. In 1791 the *Journal des Luxus und der Moden*, published by Friedrich Justin Bertuch (1747–1822), a friend of Goethe, and a liberal-minded and able businessman, described a recent arrival from London.

> A new and pretty invention is the so-called English doll which we have lately received from London. It is properly a toy for little girls, but is so pleasing and tasteful that mothers and grown women will likely also want to play with it, the more since good or bad taste in dress or coiffure can be observed and, so to speak, studied. The doll represents a young female figure, painted on cardboard, and cut out. It is about eight inches high, has simply dressed hair, and wears underclothing and corset. With it go six complete sets of tastefully coloured, cut-out dresses and coiffures, which means summer—and winter—clothing, complete dresses and négligés, caracos, chemises, furs, hats, bonnets, poufs, etc. Each dress and hat is made in such a way that the doll can easily be dressed in it, giving a fully dressed or décolleté effect while the dress fits perfectly in either case. Hat or bonnet can be adjusted freely to be pulled over the face or set back. They can be put straight or at an angle, suiting the hairstyle in a tasteful manner or otherwise. In short: dress and coiffure can be varied, and by trying, each given its particular 'air'. This dressing and undressing, being able to set up and change again, makes for the uniqueness of the English doll. One might obtain even more changes by having some extra dresses designed and painted. The whole is packed in a neat paper envelope, and can easily be carried in portfolio or working bag, for amusement at parties and for children.

All through the nineteenth century the paper doll was a popular cheap toy, and some forms of it are still current. There was a tendency at times to give this doll a portrait likeness, representing, for example, the famous Swedish singer Jenny Lind, or the dancers Maria Taglioni and Fanny Elssler. Between 1846 and 1925 the

116 Wax dolls. Re-dressed (?). France, 18th century

117 Figurine representing an old woman. Wooden jointed body, painted wooden head. With dog automaton. Germany, c. 1800

118 Dolls. All wood, blown glass eyes. Britain, first half of the 18th century

Neuruppin printers Kühn produced at least one hundred and ten different sheets of dolls, a supplement to their sheets of soldiers. The Vienna publisher H. F. Müller issued after 1810 a book, *Isabellens Verwandlungen oder das Mädchen in sechs Gestalten. Ein unterhaltendes Bilderbuch für Mädchen mit sieben kolorierten beweglichen Kupfern.* As a male counterpart there appeared with the same publisher *August's Verwandlungen oder der Knabe in sechs Gestalten* (as a beau, geologist, hermit, lancer, minnesinger, and gardener). About 1840 there followed, packed in a cardboard box, *Die Aussteuer der Puppe*, with a doll coloured on both sides and fully coloured dresses into which the doll could be slipped. Other publishers—J. and M. Trentsensky in Vienna, and Barth of Mariahilf—also issued simpler sheets, coloured, or uncoloured for painting at home. These figures could also be cut out. In the nineteenth century Britain produced dolls' furniture on sheets.

The bodies of dolls in the seventeenth and eighteenth centuries were usually made of wood with arms and legs occasionally made of kid (No. 108). Visible parts, face, neck, hands and legs from the calf down, were painted. Knees, elbows, and later shoulders, hips, and hands were movable by simple joints. To begin with wooden pins held arms and legs in place. They were fastened through the shoulder and the pelvis.

From the middle of the eighteenth century the glass-makers of Lauscha near Sonneberg provided coloured eyes of blown glass. Pasted-on wigs, mostly of natural hair, improved the life-like appearance of the doll, in spite of its still fairly stiff body. Wax proved to be a good material for the shaping of head, neck, and bust. It was poured into moulds, and could be coloured. In the eighteenth century British doll makers in particular favoured wax which up to the last decades of the nineteenth century greatly influenced doll production (Nos. 119, 121, 124, 129, 149–151). At first wax heads seemed rather monotonous compared to wooden heads. After 1850 the Montanari family of doll makers brought them to perfection.

In the eighteenth century some heads and busts were made of porcelain, a material typical of the Rococo period. Arms, legs, hands and feet were also made of this fine highly glazed material. Hard porcelain for the making of dolls' heads was later replaced by bisque and parian ware (Nos. 62, 64, 65, 160, 165, 166). Heads of these materials were unglazed, fired, but not coloured. In spite of the difference made by the glazing of porcelain, many of the finer bisque heads are just as excellent.

Following the expensively equipped fashion doll, the dress of the toy doll, particularly during the Rococo period, became more sumptuous and complicated. These dolls can supply important information on the history of costume. Together with elegant clothing, intricate hairstyles, and every detail of fashion copied from the wear of the upper classes, dolls were often given equally precious accessories. Fine ladies with

120

119 *Wax-headed doll. Britain, 1761*

120 *Wax-headed dolls dressed as Dominican nuns. Southern Germany, 18th century*

lackeys in coaches, sedan-chairs, or sledges have survived in great numbers (Nos. 114, 126, 128).

Museums tend to show only those dolls whose costume is outstanding, thus suppressing the simple fact that not all dolls were resplendent. Not all children could afford such dolls, and they had to be content with whatever their parents' or their own ingenuity could provide. Unfortunately, this type of toy is the least likely to survive. Once the child's affection has outgrown it, it reverts to what it was originally, a coarsely made doll, a knot of wood tied up in a rag that looked like a doll, perhaps even just a bundle of rags alone. Out it goes to the rubbish heap.

Society is too late in realizing what social treasures were lost when these pathetic playthings fell victims to those equally devastating forces, the child's natural growth, his parents' ideas of what is a correct plaything, or just the ever-present menace of spring cleaning.

Doll manufacturers strove hard to make the creations of their skill as life-like as possible (Nos. 110, 117).

Legends from antiquity and from medieval sources tell of machines which tried to imitate man. It is reported of Albertus Magnus (died 1280) that in the early thirteenth century he spent three decades making a servant of leather, wood, and brass. This servant could open doors and perform other useful services. It was left, however, to the eighteenth century to produce the man-like automaton of both artistic and technical perfection, for the rational eighteenth century saw in man a machine, working according to definite laws. It was hoped in time to describe the human body with the help of the exact sciences on the basis of natural-historical thought. This spirit of inquiry coupled with the pleasure in mechanical skill led to the appearance of a vast number of automated figures. Their modern successors are poetry-writing machines, computers which draw, electronically directed anthropomorphic machines, robots, and electronic brains. In the artistic and literary world there are magical and mystical figures—such as Pygmalion's statue brought to life by the goddess Venus, and the Golem, a clay figure given life by Rabbi Löw of Prague. Biology brings to mind the Homunculus formula of Paracelsus and the very latest experiments to create life artificially in a test tube. Attempts to make life-like dolls with the help of mechanical devices were continued in the nineteenth century by ingenious discoveries, copied often in the twentieth century and adapted to new technical possibilities, then presented as something completely new. When Jacques de Vau-

121 *Wax doll. Blown glass eyes. Britain, 18th century*
122 *Figurine dressed as bride; probably originally crib figure. Wooden head, hands, and feet, straw-stuffed body. Southern Germany, first half of the 18th century*
123 *Wax-headed doll. Tow-stuffed body, terra cotta hands, wooden legs, plaited hairstyle of flax. Germany, 18th century*
124 *Wax-headed dolls. Natural hair. Lady and train-bearer wearing full court dress. Britain, 1757–58*
125 *Wax-headed doll. Bridal costume. Britain, mid-18th century*

121

canson built and exhibited his famous automatons in Paris, between 1738 and 1741, a new era had begun, the era of the mechanic and the android, the automaton truly similar to man. His Flute Player 1.65 m high could play twelve pieces, producing sound by blowing and moving his lips, tongue, and fingers. Another figurine played with one hand some twenty dances on the shepherd's pipe, while playing the drum with the other hand. Friedrich von Knaus in Vienna built in 1764 a writing automaton, and between 1778 and 1780 Kempelen, Kratzenstein, and Mical built their speaking version. The androids of the Jaquet-Droz family reached the greatest fame. They are now housed in the *Musée de Neuchâtel.* The father, Pierre, a theologian, together with Jean Frédéric Leschot, built after 1770 the Writing Boy. This figure is some 70 cm high. The boy sits at a desk, dips his quill into the inkpot, moves the head slowly when writing, and even follows the stroke of the pen with his eyes, and twice shakes his quill to avoid making ink-blots. In 1773 a son, Henri Louis, a watchmaker, constructed the Draftsman which produced astonishingly complicated pencil drawings—such as a cupid with a butterfly, a portrait, or a little dog. There was also the Piano Player which appears to be breathing while playing. She presses the keys down with her fingers, turns her head, bends forward towards the music, and after playing her piece she bows and looks at the audience. Leschot again collaborated in the making of these automatons. It is not possible here to mention all the important achievements in this field right up to the present. But an example from the production belt is the children's orchestra which sprang from the inside of a walking giant robot at the opening of the World Exhibition at Osaka in 1970. Walt Disney, doyen of cartoon film-makers and the creator of Disneyland, showed at the New York World Fair of 1964 a life-size figurine of the American president, Abraham Lincoln. It was electronically directed, and according to the prospectus could take up forty-eight different poses and perform seventeen movements of head and face. This figure also made long tape-recorded speeches. We agree with Disney's biographer, Richard Schickel, that "here the dehumanization of art has been completely achieved, achieved by a man, deluded, who all the time, and with a good conscience, searched for a more human kind of art."

Doll production, especially in the nineteenth century, received inspiration from the masterpieces of mechanical skill of the eighteenth. Manufacturers in Paris, Italy,

126 Wax-headed figurine. Hands and lower part of body, wood; tow-stuffed torso, velvet and brocade dress. Germany, 18th century

127 Carnival souvenir dolls. Straw, partly coloured, dressed in cloth. Venice, 18th (?) century

128 Doll. Lady with sedan-chair. Wooden body. Germany, 1770–80

127

and Sonneberg produced some splendid mechanical dolls (No. 174).

The making of mechanical devices of the sixteenth, seventeenth, and eighteenth centuries also influenced folk art. With cribs from the Erzgebirge, for example, movement became an essential part. Present-day woodworkers make Christmas Mountains, Mines, and other models in different materials, often with several hundred wooden figures that can be set in motion. Since the fourteenth century mechanization has developed increasingly in the Erzgebirge's mining industry, and the craftsman has transferred this trend to his own creations, enjoying the mechanical 'life' of his figurines. Kept as a trade secret, there is hidden inside many a 'mountain' a complicated mechanism, a tangle of wires and threads, rails, levers, and shafts, driven by an electric motor or by clockwork, or at one time also by water or steam. Miners hammer in shafts, wood cutters saw and chop, children go sledging, boatmen row, huntsmen shoot, and even angels float in mid-air.

5 *The World of the Doll in the Nineteenth Century*

The variety of dolls and the steady technical improvement at the end of the eighteenth and during the nineteenth centuries are first hinted at in the *Magazin* of Hieronymus Bestelmeier, the Nuremberg merchant, which appeared in about 1800 in nine parts, containing 1400 drawings of objects with descriptions and prices, of which only a third, however, were toys. Bestelmeier was one of many firms which soon produced their own sample-books, even in colour (No. 139). He offered among other things:

> The English Fashion Doll (No. 218), A Woman pedlar, with movable head (No. 227), A jointed Doll, finely worked (No. 281), A mechanical Doll, dressed after the latest fashion (No. 316), A Dealer in Fashion Goods, with boutique (No. 357), Wedding present, baby made of wax (No. 527), Speaking machine, a life-size 'Sultana', splendidly dressed, mechanical (No. 555), A Nurse-maid, with Child and baby cart (No. 712).

About 1840 the catalogue of a Sonneberg firm listed as toys for girls:

> Various dolls, ordinary dolls, rattle- and dancing-dolls, 5 packs of naked dolls (they were apparently not sold singly), beautiful little maids, men's and women's faces, little cabinets, sewing boxes, small tables and chairs, standing mirrors, small carriages for dolls, desks, bird's cages, cushions for bone-lace weaving, butter churns, coffee cups, storks, simple and clattering.

During the nineteenth century the subject of the doll often turns up in children's books, frequently with moralizing and didactic tendencies. The *Hülfsbuch zur kräftigen Beförderung des Frohsinns, der Unterhaltung, Gesundheit und Heiterkeit guter Knaben und Mädchen*, which Robert Seifer published early in the century under the title *Kinderspiele* with Friese of Pirna, Germany, recommends as an entertaining game the tossing of dolls in linen table-cloths or sheets. The appendix to the book gives 54 "good rules and cautions regarding the games and pranks of children". Under No. 2 it states: "Games for boys are often very unsuitable for girls, and not taking into account propriety etc. they may even become indecent and harmful. Therefore do only play at what is suitable for you as a boy or girl." The author advises modesty in everything, and his versified rule, No. 34, reads:

> Ne'er shall I ask
> For pleasures vainly
> Which do not come within my reach.
> Let luxury never deceive me
> Though bright and tempting it may be.
> No, thinking modest, living plainly,
> Alone is joy pure and divine.

129 *Wax doll dressed in 'Langtry' tennis jersey. Britain. Doll c. 1869, probably re-dressed, as 'Langtry' was not known before 1881*

130 *Doll. Silk damask dress with train. Britain, 1883*

In the same vein Johann Carl August Musäus (1735 to 1787) published his *Moralische Kinderklapper* (Gotha, 1780), stimulated by Mouget's *Hochets Moraux*. In his story *Die Puppe*, about the naughty Patchen Philippinchen, he expressed the then fashionable sentiment that a doll may sometimes have a more beneficial influence than even a governess.

Through the work of the Montanari family, wax dolls achieved a very high standard of quality, attracting much notice, particularly at the Great Exhibition in London in 1851. A report about this exhibition states:

> The only worthwhile show of wax dolls was that of Augusta Montanari. She received a medal... The show consisted of dolls representing all ages, from infancy to womanhood, arranged in several family groups, with suitable model furniture. These dolls have eyelashes, eyebrows, and hair separately inserted in the wax, and are in other respects, too, modelled with life-like truthfulness. The dolls' ages and their stations in life are reflected in the variety of their expressions.

The comparatively high price of the dolls is explained by the amount of labour involved in the special method of fixing their hair, which required much skill and patience. The graceful beauties of the Victorian era served as models for this type of doll. The collection also popularized wax baby dolls which English manufacturers had started producing in the eighteen-twenties.

The making of wax dolls also flourished in France and Germany though technical processes varied widely. About 1875 Sonneberg produced the so-called *Staatsdame* (No. 96), a great lady in sumptuous clothes with a long train and jewellery, dressed for a court occasion. This particular wax doll was about 50 cm high, and being expensive did not remain in circulation for very long. Emmy Lehmann, the doll expert at the Deutsches Spielzeugmuseum in Sonneberg, has this to say about the background of the *Staatsdame*:

> At that time the doll cost a thaler, and only wealthy parents could afford to give it to their daughters. To working families a thaler was half a week's wages. Many of the well-to-do businessmen of that period who supplied goods to the Courts obtained the title *Hoflieferant* (by appointment—supplier to the Court) which could be bought. It was the custom to have a Court Ball annually on the Emperor's birthday, and to it the *Hoflieferant* was invited. This ball was the most important event in the life of his wife and daughters, particularly because of the exciting purchase of Court dress. The Sonneberg doll dressed in similar style was for that reason called *Staatsdame*.

The wax doll as a toy doll only lasted until the turn of the century.

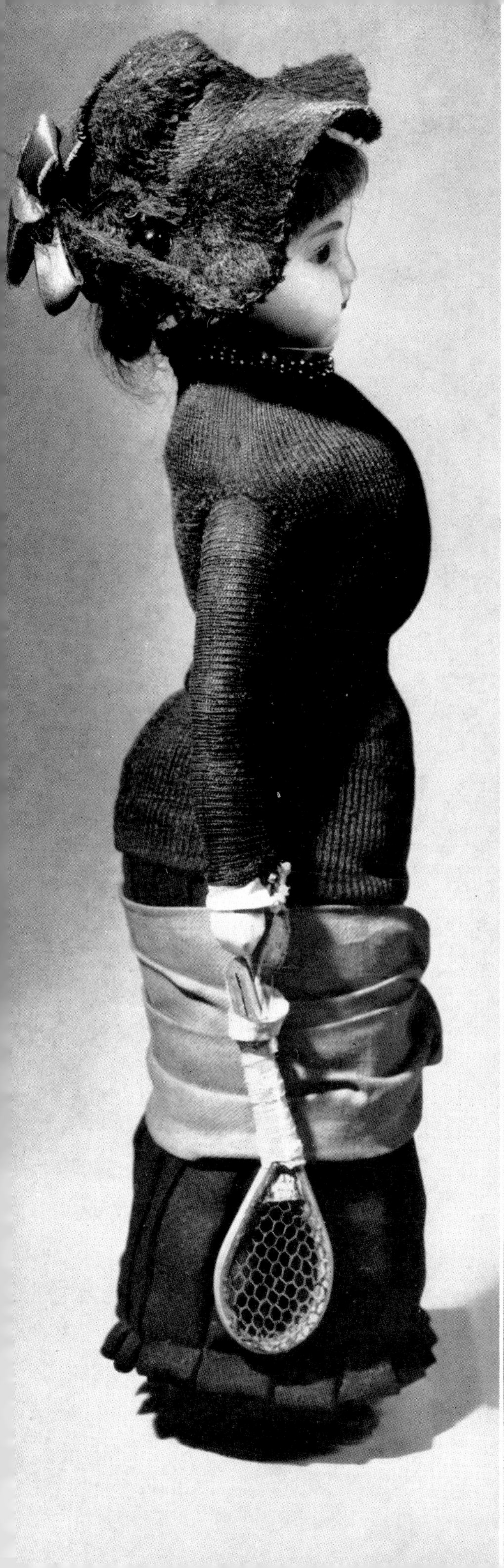

131 *Paper doll with trousseau. Vienna, 1840–45*

132 *Lumber-room of a doll's house, constructed with old toys. Germany, c. 1960*

133 *Salon of a doll's house. Basle, 1872*

In the nineteenth century dolls' heads made of porcelain with carefully modelled coiffure, mostly of black hair, were very popular. In contrast to them blonde heads were generally made of parian ware. Bisque heads—a German invention—are considered the most beautiful. Their reputation was made mainly between 1844 and the end of the nineteenth century by the Jumeau family. Strikingly expressive eyes, made by a vitrifying process, are a special feature of the Jumeau heads. In 1860 the family had a swivel neck patented for their dolls, which they made in fourteen sizes up to one metre high. The bodies, originally made of a wire frame covered with leather, were later carved of wood and put together at the joints. Eventually they were made more efficiently and much more cheaply from a mixture that could be poured into moulds and painted. In this process all joints had to be moulded separately before being joined with copper wire. The *Bébé Jumeau* attracted much attention at the Paris World Exhibition of 1855, as did later on Negro dolls with bisque heads.

As counterpart to the 'Penny wooden' there existed into the early decades of the twentieth century the cheap *Teetassen-Püppchen*, made of porcelain or bisque ware, also called 'frozen Charlottes' because of their rigid stance. These pretty dolls, some 10 to 30 cm high, were made in one piece.

Other forms of moulded dolls were those of papier-mâché or moulded paper. This technique, which came to Germany via France from China, was immediately adopted by the toy-makers, having originally been employed by such diverse tradesmen as joiners, cabinet-makers, and even coach-builders.

In the second half of the fifteenth century moulded paper was used by sculptors as material for reliefs (in the Netherlands and Alsace), and in the eighteenth century busts and statues were made of it (Jean-Antoine Houdon).

The toy industries of Sonneberg and the Meiningen uplands greatly benefited from its introduction. The new material could be easily moulded in plaster of Paris or sulphur moulds. Workshops began using it between 1806 and 1810 in place of bread dough for the finishing of wooden heads. It consisted of pulped paper, rags, rye flour, pumice stone, and kaolin, and was worked while moist, becoming hard and stiff when dry. It could be made resistant to damp by varnishing with linseed oil. The much-simplified method of production led in the Sonneberg region to further technical differentiations in this kind of work, and to the consolidation of a new large group of workers whose poor pay belongs to the darkest chapters in the history of toy-making. The almost mechanical work of the moulder required so little skill or training that toy-makers could secure their day-to-day existence only by having their many children work as well. Agents, who had to stay efficient to supply the export market, made use of every rise in demand at the expense of their workmen and families, since only by paying low wages could the dolls be produced cheaply enough to sell. In spite of all this these dolls had neatly sewn, flexible bodies, which were stuffed, and

132

finely painted heads of papier-mâché. France, the country leading in the export of dolls, bought Sonneberg dolls which were then dressed in French workshops and distributed throughout the world as *Parisienne*. In America a German immigrant, Ludwig Greiner, of Philadelphia, had the making of dolls' heads of papier-mâché patented.

In his tale *The Doll*, the Danish writer, Martin Andersen-Nexö (1869–1954) has described realistically the life of doll-makers in the Thuringian Forest. In the years before the First World War he walked extensively in this pleasant region of Germany, and learnt about the hardships of peasants, forest workers, and workers in the home industries.

During the nineteenth century further progress was made in the manufacture of mechanical dolls, particularly musical dolls driven by clockwork of German or Swiss origin. From 1820 there were speaking dolls ("For six francs I move my eyes and head, and for ten francs I say 'Papa' and 'Mama'") and in about 1878 these dolls were further developed by Thomas Edison as 'phonograph dolls' with changeable records. There were walking dolls which could move forward and backward and sideways; some could even dance the polka. The most famous of the early walking dolls was the Autoperipatetikos, patented in the United States by Joseph Lyon in 1862. There were also many other dolls with different actions, varying from the drinking dolls made by the Thuringian doll-maker Rudolf Steiner, to swimming dolls with natural swimming movements. An advertisement for mechanical toys in *New York Fashions* in the year 1877 gives an impression of the great variety available:

> The dolls imported from Paris are the finest ever. They are beautiful to look at, correctly dressed, with perfect movements. One of these represents a vine grower in peasant costume. When this doll is wound up, it walks across the floor for a few minutes, pushing a cart loaded with grapes. It holds a basket

133

in one hand, and has another basket on its back. The pretty dairy maid quickly moves about the room, shaking her head and stroking her cow which chews the cud lowing contentedly. A drunk muleteer lifts a flask to his mouth with one hand while guiding his animal with the other. A murderously inclined Zouave raves madly, firing his pistol right and left.

Again and again it is the French doll that performs the most astonishing movements. Typical of this is the 'rose doll' (about 1860) which rises by clockwork from amid the petals of the flower, then withdraws again with the petals closing over her. Another doll of the same period stands at her dressing table using her mirror and puff, while others again blow soap bubbles from a tube, and there are smoking boys and conjurers with their cups and balls.

The Sonneberg manufacturer Edmund Lindner saw at the Great Exhibition in London in 1851 the most life-like Japanese and Chinese papier-mâché baby dolls with moveable joints and the Sonneberg *Gelenktäufling*, the first Sonneberg baby doll, was made after this pattern (No. 148). The papier-mâché head was delicately painted, dipped in wax, and finally softened with wheat powder. This method gave it the natural colouring of a baby. Fine strokes of the brush at the temples and the nape of the neck hinted at the first downy hair. The baby doll's shining blue eyes were made of glass, and its dangling limbs were covered by a daintily sewn linen vest. A little crocheted bonnet completed the attractive appearance. Interest in this type of baby doll was not to last long even though it was exported to Britain, France, and other countries. It was temporarily eclipsed by the 'lady' doll, only to return in the twentieth century, brought back to favour by the growing interest in childhood.

About 1880 Heinrich Stier of Sonneberg introduced the cup-and-ball-joint doll which had long been common as the painter's lay figure, and soon after this came opening and shutting eyes, worked by a lead counterweight.

134 Dolls. Porcelain heads. Sonneberg, first half of the 19th century

Emmy Lehmann writes:

This doll became world famous. Its good qualities were praised in many languages; the natural movements of its limbs and head, the well-proportioned child-like shape and the possibility to wash and replace every part. The bisque head was made by the million from the same mould, but the many and varied coiffures of human hair or mohair allowed of pleasing variations. This doll could be dressed and undressed, it could be put to bed and to sleep, and only when it was lifted again did its eyes open. Underwear and dresses were carefully made from good patterns. It was fun playing with these dolls, and making extra clothes for them.

In the nineteenth century, too, dolls reflected the social set-up of the day. The doctor (No. 135), the coachman (No. 136), the tennis-playing young lady (No. 129), the fashionable lady dressed in lace with an elegant bag and extravagant hat, parasol, feather boa, and cashmere shawl, all of these were popular figures in the middleclass nursery.

A new nineteenth century type of doll was the portrait doll. Its forerunners were the large, 'true to life' wax figures to which Madame Tussaud and others owe their fame. Queen Victoria (1819–1901) was the first to have great numbers of portrait dolls made from her likeness. A representation of her in the splendid embroidered coronation robes of 1837 was particularly popular (No. 88). Later on the whole Royal Family—the Queen had nine children—became a constant source of inspiration for portrait dolls (No. 138). In America George Washington (1732–99), President Thomas Jefferson (1743–1826), Abraham Lincoln (1809–65), and their wives were favourite models. Jenny Lind, the celebrated 'Swedish Nightingale', and the dancer Lola Montez (No. 170), whose affair with King Ludwig I of Bavaria contributed to the tumbling of his throne in the 1848 revolution, were also made as portrait dolls. In the twentieth century in Britain and the United States public figures, such as Queen Elizabeth II, Prince Charles and Princess Anne, and Jacqueline Kennedy (Onassis) continue to be notable models.

Queen Victoria has also another and quite different significance in the history of dolls of the nineteenth century. She began to collect dolls as a young princess, old fashioned wooden ones, though in Britain at that time wax and parian ware dolls were much more popular. Her collection became world famous. Of her 132 dolls, which were about $3^{1}/_{2}$ to 8 in. high, the princess dressed 32 herself, with the aid of her governess, Baroness Lehzen. Nearly all these dolls represent well-known artists in the costumes of their parts from her favourite operas and ballets. She herself made a catalogue for the collection with detailed notes about the players and their parts. Among the dolls are also a few ladies-in-waiting and eight gentlemen. The elegant costumes minimize the stiffness of the 'Penny woodens', and raise the dolls above their original function of toys in the Royal Household to documents of social history.

135 *Character doll representing a doctor. Wooden body, movable limbs, hair and beard of fur. U.S.A., c. 1880*

136 *Bisque-headed doll representing coachman. Arms and legs porcelain, soft body. England, c. 1860*

137 *Doll with chaise. Papier-mâché head, leather body, wooden horse covered with foal skin. Christiania (Oslo), 1830–40*

138 *Wax portrait dolls. The Princess Royal and Prince Albert Edward. Britain, c. 1844*

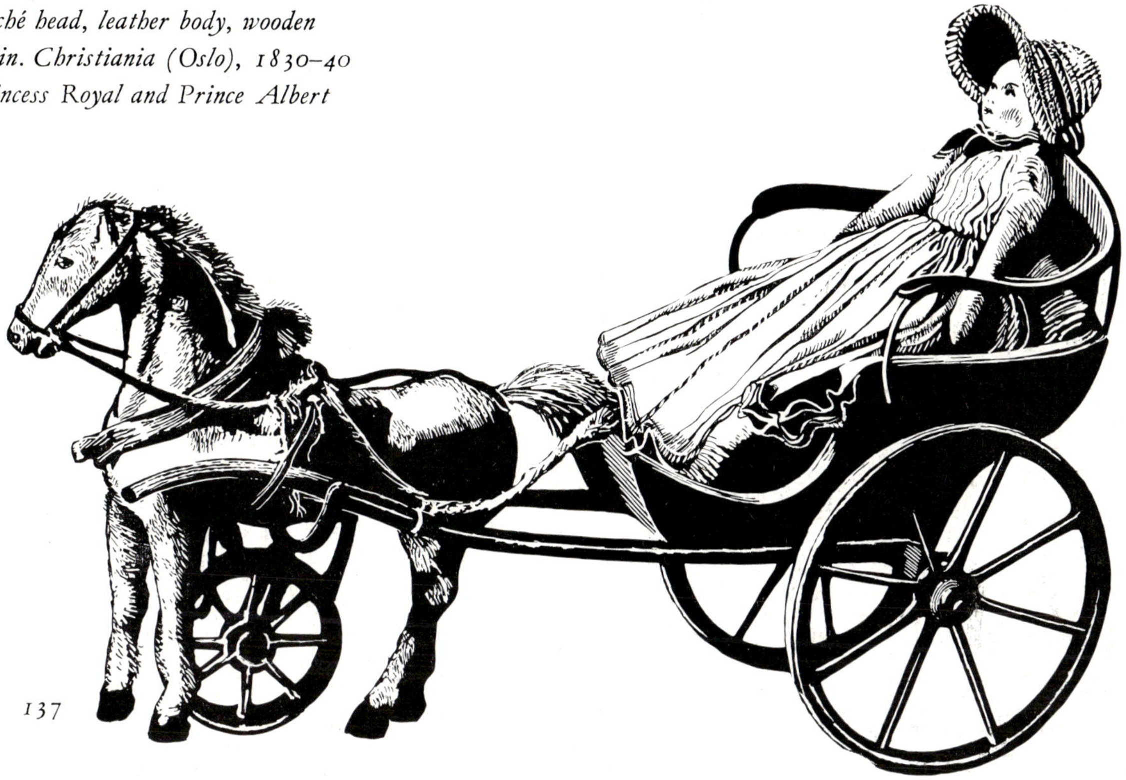

137

138

139

139 *From a Sonneberg sample-book. Germany, 1801*
140–143 *Dolls' heads. Germany, 19th century*
144 *Doll's head. Porcelain. Sonneberg, 19th century*
145 *Doll's head. Porcelain. Sonneberg, 19th century*
146 *Doll's head. Porcelain, leather body and limbs. Germany, 19th century*
147 *Doll's head. Porcelain. Sonneberg, 19th century*
148 *Baby doll. Sonneberg, 1855*
149 *Wax-headed baby doll. Britain, mid-19th century*
150 *Wax-headed baby doll. Cloth body. Britain, 1870*
151 *Wax doll. Boy dressed in wool jersey suit. Britain, c. 1865–70*
152 *Doll. Leather hands. Germany, 19th century*
153 *Wax doll. Straw hat. Norway, 19th century*
154 *Wax-headed portrait doll. England, 19th century*
155 Biedermeier *doll. Head and hairstyle partly papier-mâché. Sonneberg (?), c. 1840*
156 *Doll. 'Flirting' eyes, face papier-mâché. Sonneberg (?), 19th century*
157 *Wax portrait doll representing Sarah Penfold (born 1856). Britain, c. 1900*
158 *Wax-headed costume doll. Sonneberg, 19th century*
159 *Doll. Sonneberg, 19th century*
160 *Bisque-headed doll. 'English style' (1885–86) smocked or 'gauged' dress. France*
161 *Doll. Sonneberg, 19th century*
162 *Wax doll. France, c. 1899*
163 *Wax-headed doll. Sonneberg, 19th century*
164 *Porcelain-headed doll. Leather body, wooden limbs,* Gretchen *(German) hairstyle. Germany, 2nd half of the 19th century*
165 *Bisque-headed doll, from the Paris Exhibition of 1878. Leather body, movable limbs. France*
166 *Bisque-headed baby boy doll. Cloth body, natural hair. Germany, c. 1870–75*
167 *Wax-headed doll, dressed as Quaker. Wooden body. Britain, mid-19th century*
168 *Bisque-headed doll. Lady with ostrich-feather hat. 'Flirting' eyes. France, c. 1910*
169 *Wax portrait doll of Queen Victoria as a young princess. Britain, c. 1850*
170 *Portrait doll of the dancer Lola Montez. Head and limbs porcelain. Germany, c. 1860*
171 *Doll. Lady at a mirror. Germany (?), c. 1800*
172 *Dolls, group. Some in the dress of Munich townspeople. Germany, 1870–85*
173 *Lay figure. Wood, natural hair. With doll's trousseau of the mid-19th century. Southern Germany, early 19th century*
174 *Doll. Automaton with musical attachment. Germany, c. 1860*
175 *Doll in Court dress. Wax. Britain, 1840–50*

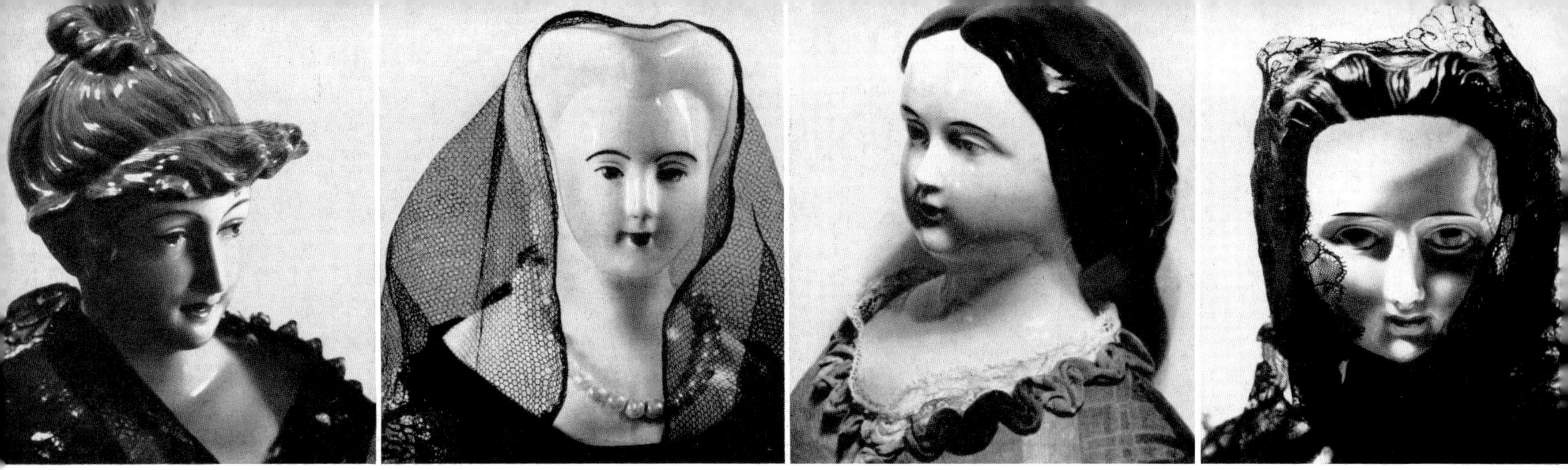

144–147

148–151
152–155

156–159

160–163
164–167

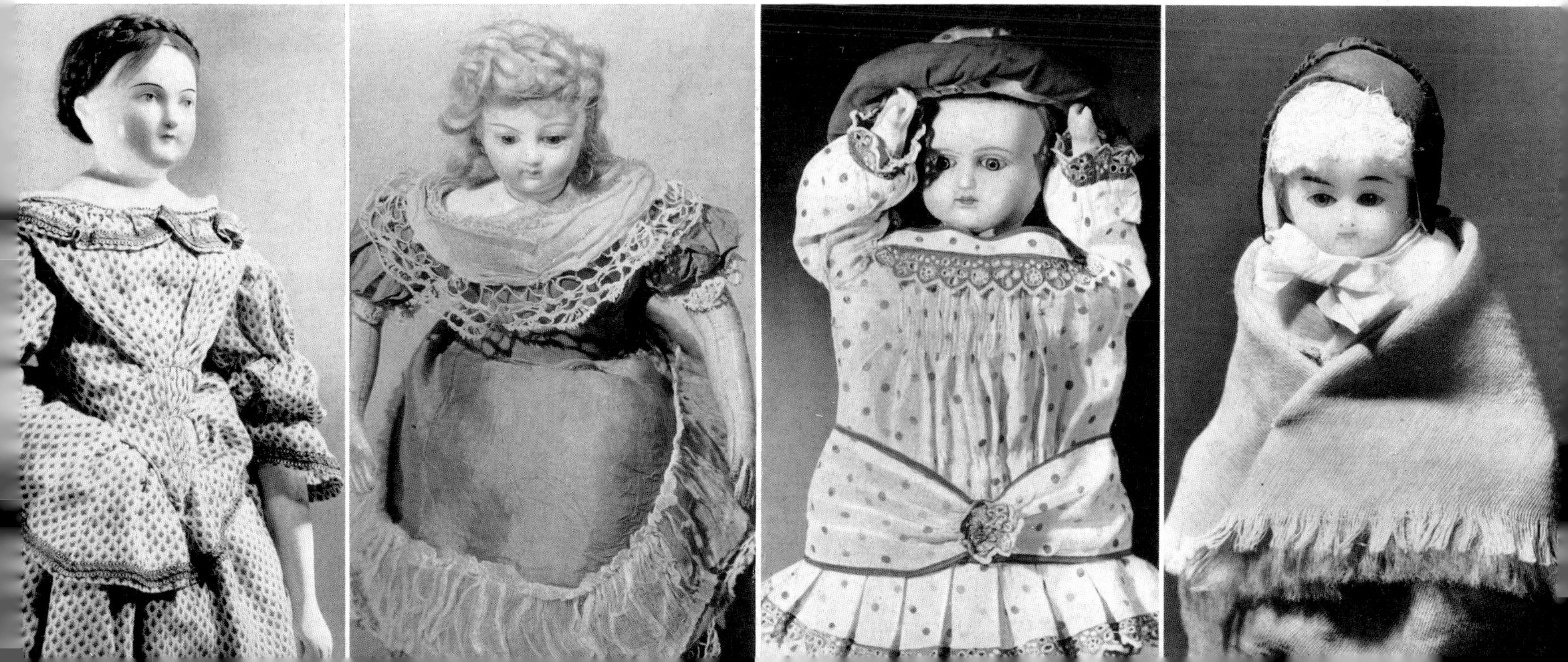

169

170

171

6 *The Doll in the Twentieth Century, the 'Century of the Child'*

German dolls dominated the world market at the beginning of the twentieth century. Two-thirds of the whole European production originated in German factories and workshops—that is, roughly half of the world's output. In the Sonneberg region alone 50 per cent of the population worked in toy production, with 25,000 people making dolls. Industrial and artistic trends in the making of dolls and toys in general were greatly influenced by social developments. The Swedish educationist and suffragette Ellen Key (1849–1926) gave a powerful impetus to educational reforms with her book *The Century of the Child* (1900). In 1840 Friedrich Froebel (1782–1852) had started the first kindergarten in Bad Blankenburg in Thuringia. Following on from the teachings of Johann Heinrich Pestalozzi (1746-1827), he recognized the value of play for the development of the child's personality, and demanded changes in the bringing up of young children. Froebel's methods soon gained ground with progressive teachers. The Italian doctor and educationist Maria Montessori (1870–1952) aimed at a pre-school education, based on the voluntary self-training of the child. Her ideas were taken up in Britain especially, but play as such was not one of her main concerns. A flood of books followed, based on the work of the pioneers, and demanding a planned training in looking at things, the enjoyment of art in schools, the development of the child's own artistic faculties, as well as recognition of the creative activity of play. There were valuable suggestions for the improvement of teaching methods. More attention was to be paid to the training of the young in folk song, play acting, gymnastics, and dance. Special congresses concerning these educational reforms were held in Dresden (1901), Weimar (1903) and Hamburg (1905).

The rise of psychology as applied to children and young people was to have a marked influence, also, on the design of toys. The ideas of the rising youth movements made themselves felt, and were, in spite of their obvious limitations a decided protest by the younger generation against a declining social order. In the course of all these efforts at solving the age-old problem of educational reform, the doll was 'discovered' for science and art. In Germany in particular, and in many other countries including France, both artists and craftsmen devoted themselves increasingly to designing toys with educational values. Naturally, in such a field, the little girl's favourite playmate, the doll, received special consideration.

In 1928 in the U.S.A. the Play School Manufacturing Company was founded, which produced toys in accordance with the new educational trends. In Britain in the twenties and thirties two teachers, Paul and Marjorie Abbatt, developed simple yet stimulating forms of toys, and opened a business in London which continues to give good advice to parents. One of the firm's catalogues says that the toys recommended are nothing special; that they are simple, only better made and more lasting than usual, and suited to the child's age. They are toys which correspond to the child's love of life, and children can turn them upside down, and use them in any way they like. The toys will neither break, nor disappoint the child.

Technical improvements in the manufacturing of dolls continued and increased during the second half of the nineteenth century. From 1840 onwards rubber was used for dolls' heads and bodies. In 1851 in the United States Nelson Goodyear obtained the first patent for dolls made of vulcanized rubber. In contrast to wax and porcelain this new material seemed ideal for children to play with.

The American Hyatt was the first to show celluloid dolls in an exhibition in 1880. Celluloid, easily washed and therefore hygienic, was a welcome newcomer, particularly for the making of baby dolls. To begin with they were made in one piece, to be given joints, glass 'sleeping' eyes, and inset teeth later on. Among American dolls, the kewpie doll, a chubby-faced baby with a little turned-up nose and a quiff of hair, was particularly attractive. Easily inflammable and brittle, celluloid soon lost in popularity. The first doll made of plastic was manufactured by the firm of Riedeler, in Königsee in the German Democratic Republic, in 1948.

A natural expression of the eyes is of great importance, since the eyes greatly influence the child's relationship to its doll. From the middle of the eighteenth century the glass-blowers of Lauscha in Thuringia made bead-like, tinted eyes, which were used for heads of papier-mâché, porcelain, and wax. There are many types of still glass eyes: black eyes (made of opalescent glass with black pupils but no irises), bead eyes with coloured irises, and millefiori eyes. For sleeping eyes the Thuringian glass-makers created the crystal eye, painted eyes which were then glazed. From 1900 transfers, too, were used for the eyes of cheap dolls. In contrast to the glass eyes made by hand, transparent plastic eyes made by an 'injection' process were mass-produced after the Second World War, when plastics became all-important. In 1961 Claude Refabert of Paris took out a patent for the plastic-made eye with a life-like look. This natural look is achieved by emphasizing the pupil and by a design which allows the white of the eyeball to be seen from all angles.

When in 1880 the Sonneberg doll-maker Stier introduced the sleeping eye worked by a lead counterweight, which in spite of new materials has remained the same in basic principle ever since, the 'naturalness' of the doll's head reached a new perfection. The doll-maker Otto Kranz, from Waltershausen in Thuringia, introduced another novelty in 1905–6: eyes that 'flirted',

176 *Wax-headed doll. Cloth body, silk dress with bead ornaments and embroidery. Britain, 1912*

177 *Wax-headed doll. Wooden body, movable bisque limbs, natural hair. Silk dress and boa. Britain, c. 1905*

the eyeballs turning from right to left. Further developments were the roguish eyes and the squinting eyes of novelty dolls and 'Aunt Sallies'. At present there seems to be a tendency to return to the painted still eye.

There exists in the German Democratic Republic a special branch of the toy industry for the making of dolls' eyes.

When at the doll exhibition at St Petersburg in 1903 many critical voices were raised against the over-elaborated French doll, a fillip was given to the reform of the toy doll. Educationists and artists protested against mass production and the fashionable doll which they said was unnatural. A child-like doll was demanded, truly representative of children, yet simplified. This reform movement spread after 1908 from Munich to Vienna, Berlin, Dresden, Karlsruhe, and other places. There were competitions, conferences, and exhibitions to promote the new ideas, and to impress doll-makers with the need for 'sensible' yet aesthetically desirable design. There was, however, a remarkable contradiction: often the new type of doll that won prizes was in its artificial naïveté not pleasing to children, and, as the critic Adolf Braig commented, extreme stylizing and over-sophistication of taste frequently led to an "almost tired beauty in dolls", and these creations remained alien to the child. Gifted artist-craftsmen endeavoured to find new ways in design, with varied success so far as child reaction was concerned. In Munich Marian Kaulitz, Lilian Frobenius, Alice Hegemann, and others succeeded in making dolls which in their simplicity and naturalness did attract children.

In 1902 a competition in France awakened the interest of artists of renown in new designs to meet powerful German rivalry in the world market, but nothing practical came of it. The decisive breakthrough was achieved by Käthe Kruse (1883–1968). In 1910 she showed in a Berlin department store *Spielzeug aus eigener Hand*, an exhibition of her first home-made dolls. They were pleasing miniature portraits of her own children and were to make the artist famous all over the world. Käthe Kruse, an actress to begin with, married to the sculptor Max Kruse and mother of seven children, was stimulated into making dolls by the demands of her eldest daughter in 1905. The first of her dolls was made of a towel, a potato, and some sand (No. 186). Under her husband's direction the dolls became more and more professional, and after 1910 conquered the hearts of all children (No. 190). "My dolls came about, especially the little dangling one, from the desire to create a feeling of holding a real child in one's arms . . . To pay attention to the emotions as an important part of the self, is, strangely enough, a new concept with regard to the doll, though on reflection it appears to be an obvious demand." These were Käthe Kruse's own words in

178 Lady at a mirror. Germany, c. 1910
179 Millinery shop. Southern Germany, 19th–20th century

179

1923. She started a workshop in Bad Kösen, near Naumburg, and in 1934 extended her work to making mannequins for shop windows. After the Second World War she moved to Donauwörth in Swabia, where her business is now carried on by her children. Her dolls obtained the *Grand Prix* at several world exhibitions.

In Italy the firm of Lenci made excellent dolls which because of their prettiness and elegant dress have become collector's items. Some of the most individual of the consciously artistic dolls, not meant for the nursery, of the first two decades of the twentieth century were Lotte Pritzel's creations, some 50 cm high; made of wire, silk, chiffon, and glittering bits and pieces, with wax heads, they are fantastic and aesthetically satisfying at the same time, displaying in their whole decor a morbid elegance, an unreal world, typical examples of the contemporary scale of values.

In a similar way Erna Pinner's dolls remain creatures of the salon, not intended for play, though they are closer to reality than Lotte Pritzel's. The Pinner dolls consist of material, painted and jointed, and are up to 150 cm high. They were given symbolic names: Sphinx, The Superior One, A Contemporary, Harmony, Pierrot, and Buddha. At a time of paralysing shortage of material in 1919 Erna Muth of Dresden produced elegant paper dolls. Their pliable wire frames allowed for a great variety of postures. Other German artist doll-makers of that period were Ursel Erbs, Wiltrud Block, Lotte Sievers-Hahn, Helma Götz, Asta Berling, and Sasha Morgenthaler (Nos. 191, 194, 198).

In the twentieth century personality dolls, great favourites with children, became successors to the portrait dolls of the last century. About 1895 the golliwog, a black man, a gay, grotesque figure from Florence Upton's book, entered the nursery. 'Golly' himself

180 Dolls on a mechanical tandem. Belonging to King George VI when a child. France (?), c. 1900

181 Bisque-headed baby doll. Home-made pillow. Germany, c. 1890–1900

180

represents a most interesting example of black dolls being favourites with white children. Sometimes this doll was dressed in an entirely fanciful costume, sometimes in an appropriate national one, but fashionable European clothes were common too. An early instance of these are Negro dolls in elegant evening dress, recorded in the southern states of America in the eighteenth century (No. 221).

Figures from Lewis Carroll's *Alice in Wonderland*, and characters from Walt Disney's films, Mickey Mouse, Minnie Mouse, Donald Duck, Huckleberry Hound, and Pinocchio followed. In America popular comedians of the screen were used as models, such as *John Bunny*, for example. In 1933–34 one and a half million *Shirley Temple* dolls were sold in the United States alone. The type lasted on the world market until 1956.

All records have been broken, however, by the teddy bear since it was first made in 1903 by the felt goods and toy factory of Margarete Steiff at Giengen in Württemberg. A political caricature by Clifford Berryman in the *Washington Post* of 1903 showed President Theodore Roosevelt (1858–1919) after a hunt in the Rocky Mountains, a small brown bear at his feet. This drawing of Teddy Roosevelt was used by what became the present American Ideal Toy Corporation for the making of their toy bears. The triumphant progress of the German bear through all of Europe began at the same time, to be followed by an English one. Steiff teddy bears were used as table decoration at the wedding of the American president's eldest daughter. Dressed as sportsmen, beaters, anglers, and hunt followers, they appeared between the dishes on the richly laid table. In the history of the firm of Steiff the years 1903–8 are known as the *Bärenjahre*, the bear years. At this time Steiff built their large factory of glass where 1400 workers raised the annual production of 240,000 toy animals of all kinds to 1.7 millions. Animals of fur, material, plush, and mohair made by Steiff, with the charac-

182 Teddy bear. Sonneberg, c. 1955

182

teristic *Knopf im Ohr* (lead seal in the ear), are known everywhere, and the Margarete Steiff company has one of the four biggest toy factories in the Federal Republic of Germany.

In the fifties television created new heroes for children, stimulating their imagination in an up-to-date manner. There were the heroes of British television, Brumas the polar bear, and Muffin the mule, as well as Torchy and Nodely. German television has its *Sandmännchen*, and Flax and Krümel, all of them friends of young viewers.

Looking at the enormous variety of modern, technically perfect dolls made of synthetic materials and plastics, but often of doubtful value, one might be inclined to speak of a crisis in the world of dolls. Stephan Hirzel, one of the best known artist-craftsmen, asks:

> What is the matter with the doll's head nowadays? If one considers 'face' as meaning the inscrutable mask of the Chinese, one has to agree that the doll lost its face a long time ago. We can take it or leave it: face or mask is a reflection of public taste. This taste is satisfied by the doll industry with a supply in cash value approximating that of the film industry. The empty smile of the film star has become an attribute of the contemporary doll also. By the old standards there is hardly any demand for character in

dolls, and it can seldom be found, and what there is, is extremely small in quantity... It need not be explained that the increasing levelling of standards in the doll's physiognomy is a necessity to maintaining currency in the world market. All the same, efforts to create better dolls will lead to a new assessment of values.

Through technical refinements, through the race for the up-to-date and the novel, and also through the conditions imposed by new materials, the modern doll is severely criticised. It is said to be a featureless mass-product, to be lacking in character, and to be wanting in everything that appeals to a child's imagination. Manufacturers are exhorted to be responsible, to create dolls which children can adopt as 'live' playmates, and things to love. Doll production therefore is not a matter only for designers and technicians together with economists. Educationists, psychologists, and even sociologists should be consulted.

A good relationship with its doll may lead the child to finding its place in society. Many a girl has the natural wish to play with dolls, and imitate her mother. Boys also treasure dolls as playmates, even if they prefer toy animals. In 1960 the International Council for Children's Play undertook research in several European countries among parents of children between the ages of three and ten, from all social classes and varying localities—city, town, village. The results of a questionnaire showed the percentage of boys and girls who played with dolls.

Country	*Boys*	*Girls*
Netherlands	51.4	96.5
German Federal Republic	78.3	97.0
Great Britain	84.0	98.0
France	72.7	100.0
Italy	67.8	98.5
Austria	84.6	98.1
Sweden	50.0	100.0

The doll may help a child in the many crises of childhood, the adaptation to school, the sadness of parting. It may help to overcome inhibitions and sublimate aggressive feelings, allowing the child to gain confidence in his fellow beings.

Looking at the international supply of dolls, including baby dolls, trends towards child-like and natural forms become more and more apparent. The 'sweet' mask-like head actually seems to be on the way out, and anatomical exactness and natural shape are taken increasingly seriously. In 1967 there was a protest in the United States by 'outraged' parents and educationists against the import of a boy baby doll of French origin, Little Brother, which, dressed only in a short vest, clearly showed its masculinity.

With the help of electro-mechanical, electronic, or electro-acoustic devices, based on nineteenth-century mechanics, toys often achieve a degree of 'naturalness'

183 *Baby doll, with soft toy animals. Soft body and limbs, composition head. Sonneberg, c. 1950*

184 *Porcelain-headed doll. Wooden hands. Thuringia (?), c. 1860–70*

185 *Doll. Painted wooden head, arms and legs, glass eyes. Germany, 1910*

186 *Early Käthe Kruse doll. Germany, c. 1913*

187 Art nouveau *doll. Germany, c. 1915*

188 Art nouveau *doll. Italy, c. 1925*

189 Art nouveau *doll. Germany, c. 1910*

190 *Käthe Kruse dolls. Germany, c. 1924*

191 *Doll. Young* Parisienne, *by Sasha Morgenthaler. Zürich, c. 1958*

192 *Doll wearing national costume. Yugoslavia, c. 1950*

193 *Cloth-finished dolls. Soviet Union, present day*

194 *Doll. Peruvian girl, by Sasha Morgenthaler. Zürich, c. 1958*

195 *Doll. Plastic. Sonneberg, 1968*

which does not allow for much play of the imagination. Speaking dolls are equipped with transistors or records. Miss Echo, 76 cm high, shown at the New York Toy Fair of 1963, has a tape inside its body on which recordings can be made and then re-tracked. From Spain comes a doll which sings to music, and moves its lips as it does so. Lilly, made by the Italian firm Sebino, speaks nine different languages. A doll produced by Riedeler in the German Democratic Republic 'masters' eighteen phrases in eight languages, and can talk for forty seconds.

The California firm of Mattel offered at the Nuremberg Toy Fair of 1970 the dancing party doll Swingy, advertised as making for gaiety, encouraging dancing, with rhythm in its body, and even carrying its own terrific record. The same firm devoted large sums of money and a two-year period of development to the creation of Dancerina, a 60-cm-high prima ballerina in a pink ballet dress. This is the first doll in the world to perform the toe dance and the pirouette. When it turns a pirouette it holds its head exactly like a real dancer. A 2.5 v. electro-motor moves the doll by means of an intricate arrangement inside the body, so that it keeps time with the music, a dance from Tchaikovsky's 'Nutcracker' Suite. The child can regulate the movements by a switch under a crown on the doll's head.

Among walking dolls, Tip Tap, made by Mattel, holds the record. Fitted with two batteries, it can walk three kilometres, and is able also to push, pull, or carry an object. The French *Blondinette* can sit down, and, controlled by two switches in its back, can walk in different directions. The firm of Schildkröt in the Federal Republic of Germany produced Stapsi, which, by adjusting the position of head and arms, can be made to walk at different speeds and in different directions.

There are dolls, too, which react to electric light or acoustic signals. The Crawling Baby is made in Japan. Driven by electricity, it obeys an acoustic order, and crawls towards the speaker, to move off at the sound of a new demand. The rat-race for novelty and superlatives in the United States has even led to a 'doll with a temperature'. The temperature rises on the intake of liquids. In 1963 a doll was produced in New York which through chemical reactions tans in the sun, to lose this tan again in the dark. The self-willed Mimi (Schildkröt) and Nanar (Bella), magnetically directed, turn away from a spoon full of spinach, but towards their milk bottle. Susi-Baby (Schildkröt) can laugh and sulk. Mattel's Weeping Doll sheds tears, cries, and wets her nappies. The French doll *Poutou* can give a smacking kiss, while Marinella by Furga of Italy and the Spanish Penito (Novo-Gama) can even belch after drinking and wetting their nappies. As a rule this type of doll is very expensive, and must therefore remain in the minority on the world market, compared with the general run of dolls. Their real value is questionable.

184–186
187–189

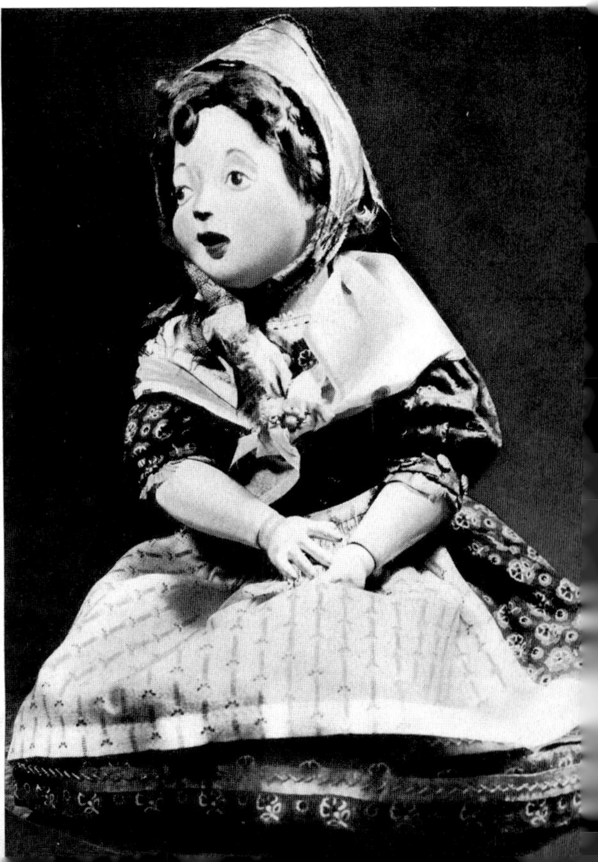

190–192
193–195

196 Soft baby doll. Sonneberg, c. 1950

The teenage-mannequin doll is intended for older girls. As a world bestseller it has been a decisive factor in the turnover of markets for the last ten years or so. To a certain limited extent, this type of doll echoes the fashion doll of the eighteenth and earlier centuries.

In 1958 the doll manufacturer Maar, from the Bavarian doll-makers' village of Mönchröden near Coburg, was inspired by Lilli, the sex symbol from the German newspaper *Bildzeitung*, to produce a new sex-appeal doll. Only a year later Mattel Inc. bought the prototype, and produced Lilli, now called Barbie. This idol, promoted by high-powered publicity, became an enormous business success. By 1968, 80 millions of the doll had been sold. Made of hard plastics, Barbie is 28 cm high, and has long legs, narrow hips, and a well-modelled bosom. The accessories of this doll, which include every aspect of fashion, earn large sums of money for the manufacturer. Meantime a whole Barbie family has appeared, comprising brothers, sisters and friends, such as Skipper, Allan, Ken, Ricky, Francis and others. In the United States alone there are more than 8500 branches of a Barbie Club, and in the Federal Republic of Germany more than 100,000 were counted in 1967. Without a doubt, Barbie was not created for children only, but was aimed also at adults. As a toy which can be added to, it brings in customers again and again, and adults like to play with this kind of doll. Edwin Faller, partner in one of the four biggest toy firms in West Germany, making plastic models and electric *Autobahnen*, has commented on this sociological phenomenon: "The turnover of toys for children cannot be increased at will; this is not taken into account sufficiently by the trade. There is a genuine demand for toys for the adult." Applied to Barbie, a West German critic analyses the facts: "Looked at carefully, Barbie is a little common, not like a girl a father would wish as a friend for his daughter. She looks more like something Father would himself like to spend a jolly evening with." The Hamburg paper *Die Zeit* reported that young men often receive the doll as a present from their girl friends. There are now several dozen teenage dolls in various countries, as a rule equipped with the creations of well-known fashion designers. Barbie became the prototype for a number of similar dolls: Tammy (U.S.A.), Cindy (Great Britain), and Tressi (U.S.A. and France). For the Italian-made Jenny the fashion designer Schubert of Rome produced over a hundred dresses, and Alexandre of Paris created the hairstyles for different occasions, such as 'At the Races', 'Cocktails at 6 p.m.', 'Picnic', and others. In 1967 Bettina was an Italian hit. She had seventeen different dresses and wigs, suited to the seasons, and for July she had a bikini and sunglasses.

At the beginning of the second quarter of the twentieth century ugly dolls were first offered in America.

197 Doll with soft toy animal. Czechoslovakia, after 1950
198 Girl doll, by Sasha Morgenthaler. Zürich, c. 1955
199 Doll. Pottery bust and arms. By Else Hoffmann, Treysa bei Kassel, Germany, c. 1968
200 Girl doll with cages. Head and hands synthetic resin. By Ute Dübbers, Heidelberg, Germany, 1968

They were designed to awaken affection in the child for people who are different and not good-looking, and so improve the child's social behaviour. This attempt has been completely forgotten, and there is now a tendency for the grotesque and ugly to be in great demand, leading directly to the anti-human shock and horror toys. The 'good-luck' Trolls, wild gnome-like fellows, a best-seller from Denmark, seem harmless compared with the sinister witches and blood-stained corpses which, like the Lindy Loonys (U.S.A.), "the abominable plastic lot—but lovable", are praised by their unscrupulous makers. A Chicago firm supplies in a box the blood-stained monsters Dracula and Frankenstein, popularized by films. Even a demonic Schicklgruber, representing Hitler, has forced an entry into the American nursery, exclaiming, "You can't win 'em all." The New York firm Aurora-Plastic offers children an electric chair in miniature, complete with offenders made of plastic. This is in the old tradition of the toy guillotine which became fashionable after the French Revolution. When Goethe wanted to give one to his son, his mother protested passionately, saying, "Dear son, whatever I can do to please you, I will do gladly and it will give me joy. But to buy such an infamous and murderous machine I will not do at any price. Were I in authority the makers of the thing would be at the pillory, and the machine itself, I should have burnt publicly by the knacker. What. To let the young play with something so outrageous, to let murder and the shedding of blood be a pastime to them, no, I'll not lend my hand to that."

From 1964 the soldier doll has been promoted in America, mainly through television. G. I. Joe is basically a counterpart to the teenage-doll. He is issued with a large number of accessories and weapons, and is supplied with kit for an infantry man, air force man, sailor, etc.

Julianne Metzger, the enthusiastic collector of toys and a passionate defender of human values in toy-making, writes:

> In earlier days animals were tormented and imprisoned in toys to amuse children, nowadays children are in danger through poorly secured electrical toys, or they are frightened by horror figures like 'Make your own Monster'. Barbie and Ken are complete, including contraceptives in miniature, and the harmless doctor set has an addition, 'the little psychoanalyst'. In that way children are exposed to experiments in trends and fashions. Hurrying on development, and the disregard for stages in this development with regard to play, can lead to damage, also in the field of technical toys . . . such damage can only be undone by careful treatment.

There is great need for children to be guarded against abuse, and for vigilance to keep their ethical and moral standards intact, allowing play to maintain its important social function.

7 *Dolls from East and South-East Asia*

In ancient Japan, with its centuries-old tradition of folk culture, its customs and religion, dolls played an important part. The first document to mention dolls is from the Heian period (A.D. 794–1192). It speaks of the symbolical-magical powers of a straw doll which in times of the plague attracted the disease to itself. It is probable that in those days dolls were also used as toys. The doll, once a cult-object and a talisman, was to become a toy, a gift of friendship, and eventually an ornament also.

Having first flourished during the seventeenth and eighteenth centuries, the artistic making of dolls has gained new ground in recent decades. Art schools, craftsmen, and gifted laymen are taking a renewed interest in doll-making. In these words J. Langner defines the role of the doll in the Far East, in his book *Ich lade Sie ein nach Kyōto:*

> Why do the Japanese with their intensity of purpose devote themselves to the cult of the doll? The doll as a fetish promises immunity. What is lifeless cannot feel pain. In the doll is concentrated the particular strength of Japan, worshipping itself. *Ningyō*, the doll, protects from illness and death . . . The doll becomes an idol and a charm against the demons in nature and in man's own soul.

The great families of Japanese dolls can be traced to the Doll Festivals, the feast of the girls on March 3rd, *Hina Matsuri*, and of the boys on May 5th, *Tango-No-Sekku*. Historical works of the Heian period give some indication of the origin of the feast. It arose from a variety of customs connected with annual and knightly feasts. The old Japanese calendar notes a religious custom at the change of the seasons, the purification ritual, when spirits and demons are driven out to fend off ill-luck. In Shintoism, the other important, ancient religion besides Buddhism, the symbol *kiyomi* means the cleaning of the heart from sins. Dolls made of coloured paper were used as scapegoats to which sins were transferred by magic. Throwing these into a river, or floating them on the water in little paper boxes, allowed one's sins, in the shape of dolls, to be swept away by the waters. At the beginning of the Genroku period (1615–1868), when during a long time of peace the arts prospered, and during the brilliant Edo period (1688–1703), dates were fixed for the two main festivals, leading to the expansion of doll-making, which was to reach a high grade of perfection.

The Girls' Doll Festival is introductory and is meant to show girls the importance of tradition in marriage and family life, as well as fostering understanding of the history and culture of their country. It is to teach reverence for the ruling house and to explain the prevailing social order. Symbolic figures represent particularly desirable qualities of character. For the Girls' Festival the best room of a Japanese house is transformed into a room for dolls. The scene, however, is not set for the girls' everyday toy dolls, but for the preciously made and handed-down dolls kept packed away during the year, according to strict family tradition. Each year

201

for three days they form the centre of the longed-for festival.

For the preparatory stage, conducted with great ceremony, the mother takes dolls and furnishings from their storing-place, and makes everything ready. To begin with a five- or six-tiered frame covered with purple crêpe is erected—the *hina dan*. The background is formed by a screen suggesting a representation of the Imperial Castle and as a rule decorated also with scenes from the capital's sights and daily life. The dolls for the festival are called *hina ningyō*.

The two chief dolls, set up with others in a certain order of rank, were at times associated with personalities admired by the people. A legend tells of a wise man, Sugawara-Michizane (A.D. 845–903) who was banished to a remote island by a jealous contemporary. The people loved him, and gave him for his wisdom, and his wife for her beauty, the honoured place of topmost couple in the little scene. After the restoration of imperial power in 1868, dolls representing the imperial couple (*dairi hina*) took top place on the stand, surrounded by their courtiers (No. 97). Close to the imperial couple stand the ministers, and on the next tier are three ladies-in-waiting (*sannin kanjo*) serving rice wine. There follow Court musicians (*go-in-bayashi*) with large and small drums and two flutes. Also belonging to the doll household are three tipplers, who with their laughing, angry, or crying expressions symbolize the three effects of drinking wine. There are, too, dolls representing famous actors of *No* plays, the classical Japanese drama, which originated in the fourteenth and fifteenth centuries from knightly musical plays, and was influenced by the ancient dances of China, Indochina, Burma, and Korea, as well as by traditional and historical material and Buddhist thought. Other dolls represent figures of the *Kabuki*, the people's theatre, a combination of drama, dance, and music, with themes taken from daily life. There are no masks in the *Kabuki* but players are heavily made up. Finally there are figures of girls, while dainty accessories, often of extreme luxury, dolls'

201 *Stick- or rod-marionettes. Silk patchwork dress. China, 20th century*

furniture, and finely made utensils and vessels complete the show. In front of the doll shrine the daughters of the house with their invited friends play, dance, and eat. The dolls, too, are served a complete miniature dinner. For the festival the mothers bake the traditional sweetmeats of sugar, rice, the leaves of the mug-wort, and plum blossom. Often the dolls are placed among branches of blossom or miniature flowering trees (No. 205).

A mirror belongs to the traditional furnishings of the festival. According to Japanese thought, the mirror is likened to a woman's soul, while the sword is like a man's soul. If the mirror is broken the marriage, too, is in pieces. In Japanese divorce is described metaphorically as the 'sorrow of the broken mirror'.

The more noble and rich a family is, the more splendid their collection of dolls. Poor families have to be contented with simple dolls cut from paper or painted on a hanging board. At a daughter's marriage the dolls go with her, and are handed down in time to her children so that they are often centuries-old heirlooms.

The great annual festival for boys has the poetic name of Festival of the Irises. It is customary for boys to take a protective bath in water in which irises have been boiled, the iris being the symbol of knighthood and of the once-privileged caste of warriors. In this the boys' festival follows knightly tradition. The tiered stand is essential also to the boys' festival. On it are placed the armed figures of historical heroes, knights, fencing masters, artists, and scientists. Outstanding among the boys' dolls are the many figurines of the Samurai, the warrior caste (Nos. 207–210). To it belonged all members of the military, from the nobility (*shogun* and *daimyos*) to the simple infantryman (No. 217). Most of the Samurai were mercenaries serving the nobility as vassals. Absolute loyalty unto death was considered the first duty of a soldier. This rule was built into the strict code of honour *(Bushido)*, which included honourable suicide *(harakiri* or *seppuku)* should Samurai

202 *Figures from the Peking Opera. China, present day*

203 *Dolls. Boys in national costume. China, early 20th century*

204 *Dolls representing hired mourner and maid-servant. Cloth with clay heads. China, present day*

duty or honour be in jeopardy. Priests of various sects also belonged to the Samurai caste, as well as all civil servants, scholars, and some doctors and artists. The Samurai were entitled to carry swords, a longer and a shorter one. The best artists and craftsmen of the country devoted themselves to their making.

To the accessories of the boys' festival belong weapons and every part of armour in miniature. In front of the house bamboo poles are erected from which flutters the 'carp banner'. Double, and sewn together from pieces of material cut in the shape of a fish and painted in bright colours, it is thought of as a symbol of health, strength, and the Samurai spirit. The air gets into the banner through a wire ring, inflating it and giving it a fish-like movement in the wind. Spiced rice cakes wrapped and baked in oak leaves are typical food for the boys' festival, as are rice dumplings cooked in bamboo leaves.

Again and again dolls are shown in the Japanese national costume, the *kimono*. It was developed from ancient Chinese Court dress of the T'ang period (A.D. 618–907). Today it is worn mainly by men, women, and children as a housegown, and is ideally suited to the Japanese style of furnishing without chairs.

The ancestors of today's Japanese toy dolls are numerous and varied. There are many kinds of pretty paper dolls *(kami hina)*, often made by children themselves. The heads are covered with composition, and painted, with bodies made of silk-covered cardboard, or folded paper (Nos. 211, 212, 241).

There are records of clay toy dolls in the seventeenth century, and their descendants exist in folk art as *fushime ningyō* to this very day. They are representations of gods of good fortune, figures from history, folktales, and legends. Collectors consider the *hakata ningyō* the finest achievement of Japanese doll-making. Modelled on dancers, actors, and other figures, they represent customs and ethnic characteristics of the Japanese.

As early as the seventeenth century, long before the material was used in Europe for dolls, papier-mâché dolls were made in Japan, particularly in the centre of the country. Now it is used only for toy animals and the *daruma* (cork-tumbler).

Among wooden dolls, those carved from the wood of the *Kiri* tree *(gosho ningyō)* are considered the finest. Subtly painted, they were in the early Edo period much sought-after ornaments for the palaces of the noblemen, and were given as presents on the birth of a child, and they served as a travel talisman for ladies. It was natural that the best makers of *gosho ningyō* settled in the old imperial city of Kyōto. There are also *saga ningyō*, wooden dolls popular until the nineteenth century. Their colours were laid down by Buddhist teachings. In contrast to the *gosho ningyō* which represent children, *saga ningyō* dolls were figures of gods, warriors, noblemen, characters from daily life, and *No* actors. The *ukiyo ningyō* originated in the eighteenth century as a figure from daily life; several of them could be assembled into scenes. Their costumes became models for the Doll Festivals. The *ukiyo ningyō* had a body made of wood or straw, with clothes pasted on singly and built up into the final costume, or a ready-sewn dress might be slipped over the doll's head. In order to look elegant and graceful these dolls had few undergarments, and their limbs were often

205 Dolls. Girls in an orange tree. Japan, present day
206 Cloth doll. Java, c. 1900

206

207 Doll. Soldier with spear and bow. Japan, 1932
208 Doll. Mounted Samurai. *Japan, present day*
209 Doll. Samurai *with sword and fan. Japan, 1932*

elongated. There has been a revival of the *kami ningyō*, also called *kimekomi ningyō*. The term *kimekomi* refers to the technique used in the dressing of these dolls, and implies the use of incisions. The bodies of the carved wooden dolls are fitted all over with grooves into which their clothes, of brocade or crêpe, are then pasted. Among the *kimekomi* are figures of children, noblemen, townspeople, *No* actors, and gods of fortune. The complicated *mitsure ningyō*, a jointed dressed doll made of wood or papier-mâché, was a life-like representation of Japanese children or *Kabuki* actors. Their style has been preserved in the *yamato ningyō*, which has no joints, and has been made since 1926, mainly as an ornament. The simple *kokeshi ningyō*, a wooden doll turned in the shape of a cylinder, with a rotating head, and painted with decorative designs, originated probably as a toy and souvenir in the mid-nineteenth century.

Today Japan is, after the U.S.A., the largest producer of toys in the world, and accounts for about a third of the world's toy production. Some 20,000 people are employed in about 2000 places making toys, 50–60 per cent of these in very small workshops or in their own homes. Characteristic of the economic structure of the industry is that producers are linked to big trading firms. Among materials used, metal and synthetics take first place, the latter particularly for the making of dolls, in which about half of the producers specialize. It is surprising that in spite of the big volume of production nearly all dolls are sold on the home market. One firm alone, the biggest among nearly 1000 makers of dolls, the Hakata Urasaki Doll Company in Fukuota City, offers 2500 different kinds of dolls.

In other East Asian countries, too, doll-making is an important branch of craftsmanship, in high repute to this very day. In the languages of China and Korea the word for doll has the same root as that for idol or fetish, actually hinting at the doll's religious significance. In China archaeologists have discovered doll-like grave figures of ceramics and wood. These figures, which sometimes appear to have movable limbs, support the assumption that toys, too, were made at an early date. Puppets and movable shadow-figures belong to the oldest tradition of the Chinese theatre. Legend tells that in the second century B.C. the chief of the Huns, Mao-tun, besieged the Emperor, Kao Tsu of the Han dynasty, in the city of Ping (Shansi). Capitulation seemed inevitable, had the Emperor not found out that Mao-tun's wife was very jealous. The Emperor then arranged for a number of pretty wooden dolls to dance on the city walls. Mao-tun's wife, fearing to lose her husband's affection, compelled him to withdraw at once.

An enchanting Chinese ink drawing on silk, ascribed to the twelfth or thirteenth century and now at the Cleveland Museum of Art, shows one hundred children at play, among them a girl nursing a doll in her arms. Simple dolls of plaited straw, made even now by mothers and children, were known in prehistoric times at sacrificial ceremonies.

Chinese craftsmanship, with its fine products of ivory, marble, soapstone, wood, metal, silk, lacquer ware, and

210 *Armed* Samurai. *Probably a satirical figure. Japan, 1932*

211 *Folded paper dolls. Heads covered with stucco and painted, bodies cloth-covered cardboard. Japan, beginning of 20th century*

211

212 *Paper dolls. Heads papier-mâché with stucco, dresses silk. Japan, 20th century*

213, 214 *Boy dolls. Japan, present day*

215 *Jointed papier-mâché doll. Covered with stucco and painted, painted eyes. Japan*

216 *Doll. Boy with stick horse, symbol of the Boys' Doll Festival. Japan, present day*

217 *Figure for the Boys' Doll Festival. Sword dancer. Japan, present day*

218 *Dolls in the classical style. Japan, present day*

219 *Dolls in historical costume. Soft body, natural hair, limbs papier-mâché. Japan, c. 1900*

220 *'Mantelpiece' doll, representing a young lady of the upper middle class of the Genroku period. Torso and limbs porcelain, silk hair. Japan, c. 1960*

212

porcelain, has done much for the artistic design of the doll in general. Excellently painted clay dolls, dainty ones on small sticks which when assembled represent a train of gods or demons (No. 201), richly dressed boy dolls (No. 203), simple dolls full of character representing figures from daily life, with soft bodies and clay heads (No. 204), the gay *daruma* of papier-mâché, and a great variety of modern dolls, their traditional costumes made of colourful materials, are examples of the subtle artistry of Chinese doll-making in the twentieth century (Nos. 81, 98).

Figures from the Peking opera, the Chinese musical theatre which originated in the nineteenth century, are popular models, too, for the doll-maker (No. 202). Subjects of this opera include historical events, heroic fights of the past, legends, folktales, and bucolic farce. The colourful costumes and highly stylized masks, characteristic for certain parts, go back to time-honoured custom familiar to the people.

In ancient India dolls were made of wood, buffalo horn, ivory, or wool. Tradition advises boys and youths who wish to gain the affection of maidens to share their play with dolls. In 1873 in Bengal a magnificent dolls' wedding was celebrated publicly. Among the crafts of India, which exports to many countries precious materials embroidered with gold and silver threads, metal work, wood carvings, and colourful lacquer ware, doll-making still plays a great part. Painted dancing dolls, hand-made in terracotta (from Assam and Bengal), figures of clay or wood from Uttar Pradesh, and brilliantly painted dolls of wood and papier-mâché from Tanjore in southern India are proof of the variety of the craftsman's skill. Dancing girls, mother and child, and women wearing the *sari* are always popular motifs. The Indian historian Susheela Rajni Patel created 1000 dolls, and in more than sixty dioramas presented events in the history of her country. In 1967 her collection was shown in the German Democratic Republic.

Burma produces wooden dolls with movable joints, painted dancing dolls, and the *daruma* of clay representing the Buddhist apostle Dharma.

8 *Souvenir Dolls - Ambassadors of Good Will*

The souvenir doll is a product of the eighteenth and nineteenth centuries which has become popular among collectors. Souvenir dolls made of straw appeared in the eighteenth century at the Venice Carnival (No. 127).

An early example of interest in other peoples' costume is to be found in *Kinderspiele*, published by Robert Seifer in Pirna, Germany, at the beginning of the nineteenth century. A fashion doll is offered in the appendix, costing sixteen *groschen*, with eight different sets of clothing. This clothing, says the advertisement, includes "the latest outfits favoured by the ladies, as well as special national costumes of other European peoples, and so providing useful information when selecting fancy dress ..." This doll therefore relates in some ways to the later 'tourist piece'.

The World Exhibitions of the nineteenth century contributed greatly to the spread of costume and souvenir dolls. At the first Exhibition in London in 1851 the Sonneberg doll-makers showed two large exhibits with hundreds of animated dolls and figures: Gulliver's Awakening in the Country of Lilliput, now in the Deutsches Spielzeugmuseum at Sonneberg, and Fête in the Rosenau, which was sold in London. At the Paris World Exhibition of 1900 a scenic display of German fairytales and a figure of Father Christmas, with a sledge drawn by two life-size stags, publicized the

221 *Negro dolls in contemporary evening dress. Southern states of the U.S.A., 18th century*
222 *Souvenir dolls. Wool. Portugal, c. 1960*
223 *Dolls,* matryoshka. *Soviet Union, c. 1965*
224 *Bride doll. Slovakia, early 20th century*
225 *Dolls wearing Breton costume. France, 20th century*
226 *Doll wearing Norwegian national costume. Norway, 20th century*

221

work of the Thuringian toy-makers. In Brussels in 1910 they received a *Grand Prix* with their exhibit Thuringian Fair, now also in the Deutsches Spielzeugmuseum.

The Chicago World Exhibition of 1893 included a collection of twenty-five dolls, models documenting the development of women's fashion in France. In the same year wives of settlers in Dutch East India presented to the then very young Queen Wilhelmina a collection of dolls which provided a survey of costumes worn in the colonies. At the World Exhibitions in Paris in 1867 and Vienna in 1873 products of folk art were included in the programme for the first time. Costumes of different countries were displayed on life-size figures, emphasizing the national element and providing research material for craftsmen and the textile industry. Peasant costumes, still worn in different regions, were taken as models. They had been 'discovered' by increasing travel and tourism, and were often considered quaint. To meet this growing interest in costume all the countries of Europe were inspired to make costume dolls. In Germany these trends coincided with the general discovery of folk art as a subject of serious study. At the turn of the century folklore associations, journals, and museums were started. Peasant furniture became fashionable, and the 'simple country way of life' was advocated by enthusiasts.

Nowadays the collecting of costume dolls is an adult hobby. This is particularly true of craftsman-made dolls, often correct in every detail, hence fairly expensive and rarely in demand as toys. As a rule they are ethnographical souvenirs only, pleasant little ornaments to bring home from a journey. A special example is the precious dolls wearing bridal dress (Nos. 224, 235, 238). Stephan Hirzel has considered the educational value of souvenir dolls, and writes:

240

239 *Figures. Slovak tinker and child pedlars from the* Striezelmarkt. *Seiffen, Erzgebirge, c. 1960*

240 *Yakut dolls. Siberia, early 20th century*

241 *Paper doll. Head papier-mâché with stucco. Japan, 20th century*

Striezelmarkt. The aging Seiffen toy-maker Max Auerbach still makes this pretty group by hand, and it is sold all over the world. Though not toys in the true sense, the little figures have proved genuinely attractive to children.

Examples from Mexico (Nos. 237, 242), Algiers (No. 247), Peru (No. 230), the Red Indians (Nos. 231, 232), and the Zulus (No. 245), show that souvenir dolls are popular outside Europe, too. Increasing export of dolls to African countries and sales of dolls there show that a demand exists on that continent, particularly for inexpensive dolls. In central Africa black dolls are in great demand, except in the highlands of Madagascar, where the inhabitants are not of African origin, but are generally light-skinned. There is a great demand for souvenir dolls, mostly made of foam rubber, ochre coloured or black, and dressed by local textile firms. With growing travel in Africa tourists are ever-ready buyers.

Souvenir dolls, in all their variety, show the ethnographical and historical characteristics of peoples, and are ideal for recalling memories of distant places. As ambassadors of goodwill they fulfil a useful function.

241

242 *Figurines. Water carriers. Probably clay. Mexico, present day*

242

Sources of Illustrations

The authors and publishers wish to thank the following museums and collections who made items available for reproduction in this book:

Bayerisches Nationalmuseum, Munich: 89, 93, 105, 106, 110, 114, 115, 117, 123, 126, 127 below, 164, 168, 185 and back of the dustjacket

Benaki-Museum, Athens: 10, 11, 23, 30

Bethnal Green Museum, London: 42, 60, 62–66, 71, 72, 77, 80, 88, 90, 94, 98, 101, 102, 104, 107, 113, 119, 124, 125, 129, 130, 136, 149–151, 154, 157, 160, 165, 167, 169, 171, 175–177, 181, 204, 207–210, 220, 227 228, front of the dustjacket

British Museum, London: 18, 20, 37–40

Collection Jan Balet, Munich: 1, 135, 186, 222, 230

Collection Peter His, Basle: 100, 111, 132, 133

Collection Sasha Morgenthaler, Zurich: 67, 191, 194, 198

Deutsches Spielzeugmuseum, Sonneberg: 8, 13, 16, 26, 32, 35, 36, 43, 48, 50 left, 51, 53, 54, 56–59, 75, 79, 81, 83, 84, 86, 87, 91, 96, 97, 99, 109, 116, 127 above, 134, 139–145, 147, 148, 152, 156, 158, 159, 161, 163, 178, 183, 187–190, 192, 193, 196, 197, 201–203, 205, 213, 214, 216–219, 224, 225, 229, 233–239, 242, 247

Germanisches Nationalmuseum, Nuremberg: 2, 50 right, 120, 128

Landesmuseum Oldenburg: 6

London Museum, London: 102 below left, 121, 138, 221

Louvre, Paris: 47

Museo Civico, Aquileia: 45

Musée d'Ethnographie, Neuchâtel: 22, 33, 49

Museum of Childhood, Edinburgh: 69

Museum für Völkerkunde, Leipzig: 7, 9, 12, 14, 15, 17, 19, 21, 24, 25, 27–29, 31, 34, 52, 61, 73, 78, 103, 206, 211, 212, 215, 240, 241 and page 6

Nordiska Museet, Stockholm: 68, 146, 226

Norsk Folkemuseum, Oslo: 137, 153;

Private collections: 41, 92, 182, 195, 223, 244

Puppentheatersammlung der Stadt München, Munich: 174

F. Radspieler & Co., Munich: 199, 200

Rijksmuseum, Amsterdam: 3

Schweizerisches Museum für Volkskunde, Basle: 70, 76, 246

Smithsonian Institution, Washington: 231

Staatliche Kunstsammlungen, Dresden: 5

Staatliche Museen zu Berlin: 44

Staatliches Museum für Völkerkunde, Munich: 245

Staatliches Museum für Volkskunst, Dresden: 95

Stadtmuseum, Munich: 112, 122, 131, 155, 166, 170, 172, 173, 179, 184

Trachtenpuppen-Museum, Neustadt bei Coburg: 74, 232

Victoria and Albert Museum, London: 82, 85, 108, 118, 162

Westminster Abbey, London: 55

Windsor Castle: 4, 180, 243

For the photos of the following illustrations we are grateful to:

Kunstarchiv Arntz, Haag/Oberbayern: 2, 3

Landesmuseum Oldenburg: 6

Staatliche Kunstsammlungen Dresden: 5

All other photos are by Claus Hansmann, Munich

Our thanks are also due to Liselotte Hansmann for help with the editorial work

Bibliography

d'Allemagne, H. R.: *Histoire des jouets* (Paris, 1902).

Alt, R.: *Die Erziehung auf frühen Stufen der Menschheitsentwicklung* (Berlin, 1956).

Alte Puppensammlung (catalogue) (Cologne, 1973).

Altes Spielzeug, in the periodical *Du,* No. 11 (Zurich, 1951).

Arbeitsausschuß Gutes Spielzeug e.V.: *Gutes Spielzeug* (Ravensburg, 1961).

Asiatische Schattenspiele (catalogue) (West Berlin, 1969).

Aslin, E.: *Toys* (Victoria and Albert Museum, London, 1967).

Bachmann, M.: *Seiffener Spielzeugschnitzer* (Leipzig, 1956).

– *Zur Geschichte der Seiffener Volkskunst,* in *Abhandlungen und Berichte des Staatlichen Museums für Völkerkunde Dresden,* Vol. 28 (Berlin, 1968).

– *Zur Entwicklung der Erzgebirgischen Holzschnitzerei,* in *Abhandlungen und Berichte des Staatlichen Museums für Völkerkunde Dresden,* Vol. 22 (Berlin, 1963).

Bachmann, M., and Langner, R.: *Berchtesgadener Volkskunst* (Leipzig, 1957).

Baird, B.: *The Art of the Puppet* (New York, 1965).

Baldet, M.: *Figurines et soldats de plomb* (Paris, 1961).

Bayer, L.: *Altes Spielzeug,* in *Kaysers Kunst- und Antiquitätenbuch,* Vol. III (Munich, 1967).

– *Das europäische Puppenhaus von 1550–1800* (part of a dissertation) (Würzburg, 1962).

Beitl, R.: *Wörterbuch der deutschen Volkskunde* (Stuttgart, 1955).

Benedict, R.: *Patterns of Culture* (London, 1935).

Benjamin, Walter: *Über Kinder, Jugend und Erziehung* (Frankfort-on-Main, 1969).

Berliner, R.: *Die Weihnachtskrippe* (Munich, 1955).

Bernhard Schultze/Ursula (catalogue) (Rotterdam, 1974).

Bestelmeier, G.-H.: *Magazin von verschiedenen Kunst- und anderen nützlichen Sachen* (Nuremberg, c. 1800).

243 *Eskimo doll. Greenland, 19th–20th century*

244 *Nine-pin dolls. By Horst-Joachim Tappert, Dresden. Germany, c. 1968*

245 *Zulu dolls. Leather. South-east Africa, beginning of 20th century*

Bilz, Helmut: *Die gesellschaftliche Stellung und soziale Lage der hausindustriellen Seiffener Spielzeugmacher im 19. und Anfang des 20. Jahrhunderts* (Seiffen, 1975).

Boehn, M. von: *Puppen* (Munich, 1929).

– *Puppenspiele* (Munich, 1929).

Boesch, H.: *Kinderleben in der deutschen Vergangenheit* (Jena, 1924).

Boguslavskaya, I.: *Russkaya glinyanaya igrushka* (Leningrad, 1975).

Böhmer, G.: *Püppchen* (Cologne, 1963).

– *Puppentheater* (Munich, 1969).

– *Die Welt des Biedermeier* (Munich, 1968).

Bolte, J.: *Zeugnisse zur Geschichte unserer Kinderspiele,* in *Zeitschrift des Vereins für Volkskunde,* Vol. 19 (1909).

Böttger, W.: *Führer durch die ostasiatischen Sammlungen* (Leipzig, 1958).

Bringmeier, M.: *Wandel der Mode im Zeitalter der Aufklärung,* in *Rheinisch-Westfälische Zeitschrift für Volkskunde,* Vol. XIII (Bonn, 1966).

Bühler, K.: *Die geistige Entwicklung des Kindes* (Jena, 1930).

Burky-Bartelink, A.: *Antiek speelgoed* (Bussum, 1973).

Catalogues of the toy mail-order firm Franz Carl Weber (Zurich, 1960–68).

Chekalov, A.: *Bäuerliche und russische Holzskulptur* (Dresden, 1967).

– *Russisches Volksspielzeug* (Dresden, 1972).

Christensen, N.: *Über das Wesen des Spiels* (Berlin, 1962).

C. M. W.: *Doll's Houses* (London, 1965).

Cohen, John: *Golem und Roboter. Über künstliche Menschen* (Frankfort-on-Main, 1968).

Coleman, Dorothy, Elizabeth and Evelyn: *The Age of Dolls* (Washington, 1965).

– *The Collector's Encyclopedia of Dolls* (New York, 1968).

Culff, R.: *The World of Toys* (London, 1969).

Daiken, L.: *World of Toys* (London, 1963).

Das Kinderbild von Meisterhand (catalogue) (Weimar, 1972).

Dickens, C.: *The Cricket on the Hearth* (Edinburgh, n. d.).

Dolls (catalogue) (London, 1969).

Dolls of Kyoto (catalogue) (Kyoto, 1970).

Dörig, J.: *Von griechischen Puppen,* in *Antike Kunst,* No. 2 (Olten, 1958).

Drost, D.: *Kunst aus Afrika* (Leipzig, 1963).

Farbe, Motiv, Funktion. Zur Malerei von Naturvölkern (catalogue) (Basle, 1969).

Forman, W. and B.: *Kunst ferner Länder* (2 vols., Prague, 1956 and 1959).

Fraser, A.: *A History of Toys* (London, 1966).

– *Dolls* (London, 1963).

Fraser, D.: *Primitive Art* (London, 1962).

Fritzsch, K. E.: *Motive des Spielzeugs nach erzgebirgischen Musterbüchern des 19. Jahrhunderts,* in *Sächsische Heimatblätter,* No. 6 (Dresden, 1965).

Fritzsch, K. E., and Bachmann, M.: *An Illustrated History of Toys* (London, 1966).

Froebel, F. W., (rendered into English by S. S. F. Fletcher and J. Welton: *Chief Writings on Education* (London, n. d.).

Gantner, Theo, and Waltraud Hartmann: *Das Spielzeugbuch* (Frankfort-on-Main/Innsbruck, 1973).

Geist, H. F., and Mahlau, A.: *Spielzeug. Eine bunte Fibel* (Leipzig, 1929).

– *Die Grundlagen der afrikanischen Kultur* (Leipzig, 1948).

Geschichte der Erziehung (by a group of authors) (Berlin, 1969).

Giesen, J.: *Europäische Kinderbilder* (Munich, 1969).

Greene, V.: *English Dolls' Houses* (London, 1955).

244

Gröber, K.: *Kinderspielzeug aus alter und neuer Zeit* (Berlin, 1927). Second edition, ed. by Juliane Metzger (Hamburg, 1965).

– *Das Puppenhaus einer deutschen Fürstin* (Königstein and Leipzig, n.d.).

Groos, K.: *Das Spiel* (Jena, 1922).

Hansmann, C. and L.: *Viel köstlich Wachsgebild* (Munich, 1959).

Hasse, Max: *Spielzeug und Spiele* (Lübeck, 1974).

Heinrichs, M.: *Stoff, Fell, Filz und Faden. Ein Puppen-Bastelheft* (Freiburg, 1966).

Hermann, F., and Germann, P.: *Beiträge zur afrikanischen Kunst* (Berlin, 1958).

Hertz, Louis H.: *The Toy Collector* (New York, 1967).

Hetzer, H.: *Kind und Jugendlicher in der Entwicklung* (Hanover, 1948).

Hildebrandt, P.: *Das Spielzeug im Leben des Kindes (Berlin,* 1904).

Hills, J.: *Das Kinderspielbild von Pieter Bruegel d. Ä.* (1560) (Vienna, 1957).

Himmelheber, H.: *Eskimokünstler* (Eisenach, 1953).

Hirschberg, W., and Janatz, A.: *Technologie und Ergologie der Naturvölker* (Mannheim, 1966).

Hirzel, S.: *Spielzeug und Spielware* (Ravensburg, 1956).

Holme, G.: *Children's Toys of Yesterday* (New York, 1932).

Huizinga, J.: *Homo Ludens* (London, 1949).

Hürlimann, B.: *Kinderbilder aus fünf Jahrhunderten europäischer Malerei* (Zürich, 1949).

Ich bin wieder da! Altes Spielzeug (catalogue) (Bremen, 1961).

Ilyin, M.: *Die russische dekorative Volkskunst* (Moscow, 1959).

Jacobs, F. G.: *A History of Dolls' Houses* (London, 1954).

Jacobs, F. G., and Faurholt, E.: *Dolls and Doll Houses* (Japan, 1967).

Jouets. Une sélection du musée de Sonneberg, R. D. A. (catalogue) (Paris, 1973/1974).

Jugend in drei Jahrtausenden der Kunst (catalogue) (Berlin, 1973).

Jünger, F. G.: *Die Spiele* (Munich, 1959).

Kagan, M.: *Vorlesungen zur marxistisch-leninistischen Ästhetik* (Berlin, 1969).

Katsube, A.: *Traditionelles Spielzeug in Japan: eine entschwindende Welt,* in *Graphis,* No. 141 (Zürich, 1969).

Kaut, H.: *Alt-Wiener Spielzeugschachtel. Wiener Kinderspielzeug aus 3 Jahrhunderten* (Vienna, 1961).

Keller, G., translated by P. B. Thomas with the collaboration of B. Q. Morgan: *A Village Romeo and Juliet* (London, 1955).

245

Kind und Spiel (catalogue) (Nuremberg, 1968).
Das Kind und seine Welt (catalogues) (Vienna 1960 and 1963).
Kinder-Kaleidoskop (catalogue) (Hagen, 1964).
König, W.: *Das Holz und Werkzeuge zu seiner Bearbeitung,* in *Katalog Arbeit und Werkzeug* (Leipzig, 1962).
Kuczynski, J. and Hoppe, R.: *Geschichte der Kinderarbeit in Deutschland* (2 vols., Berlin, 1958).
Kulischer, J.: *Allgemeine Wirtschaftsgeschichte des Mittelalters und der Neuzeit* (2 vols., Berlin, 1954).
Kunst als Spiel – Spiel als Kunst (catalogue) (Recklingshausen, 1969).
Kybalová, L., Herbenová, O., and Lamarová, E.: *Das große Bilderlexikon der Mode* (Gütersloh, 1966).
Leber, W.: *Die Puppenstadt Mon Plaisir* (Leipzig, 1965).
Lehmann, E.: *Die Puppe im Wandel der Zeiten* (Sonneberg, 1957).
– *Die Sonneberger Puppenmacher,* in *Deutsches Jahrbuch für Volkskunde,* Vol. IV, Part II (Berlin, 1958).
Lips, J.: *Vom Ursprung der Dinge* (Leipzig, 1961).
Makarenko, A. S.: *Vorträge über Kindererziehung* (Berlin, 1951).
– *Ausgewählte Pädagogische Schriften* (Berlin, 1952).
Man and his World (catalogue) (Montreal, 1967).
Marwitz, C. von der: *Spielzeug aus Frankfurter Familienbesitz* (Frankfort-on-Main, 1965).
– *Der kleinen Kinder Zeitvertreib* (Darmstadt, 1967).
Maurice, Klaus: *Von Uhren und Automaten* (Munich, 1968).
Mechanisches Spiel und Theater (catalogue) (Munich, 1972).
Metzger, J.: *Nie mehr Langeweile. 12 Rezepte für Ärzte und Eltern kranker Kinder* (Karlsruhe, n.d.).
– *Spielsachen richtig kaufen und selber machen* (Lahr, 1962).
– Spielzeug damals, heute, anderswo (Frankfort-on-Main, West Berlin, 1964).
– *Spielzeug dreimal täglich,* in *Medizinischer Wochenspiegel,* No. 6 (Darmstadt, 1966).
Metzger, J., and Duis, H.: *Die Spielzeugbude* (Munich, 1965).
Mozhaeva, E.: *Matryoshka* (Moscow, 1969).
Murray, P.: *Toys* (London, 1968).
Müller, W.: *Japanisches Mädchen- und Knabenfest,* in *Zeitschrift für Ethnologie,* Vol. 43 (Berlin, 1911).
Nachdenken über Spielsachen (catalogue) (Nuremberg, 1973).
Neumann, K.: *Spielzeugland Sonneberg* (Gotha, 1939).
Niki de Saint Phalle (catalogue) (Hanover, 1969).
Noble, J.: *Dolls* (London, 1967).
Norwegische Volkskunst (catalogue) (Hamburg, 1962).

246 Doll wearing Macedonian costume. Before 1900

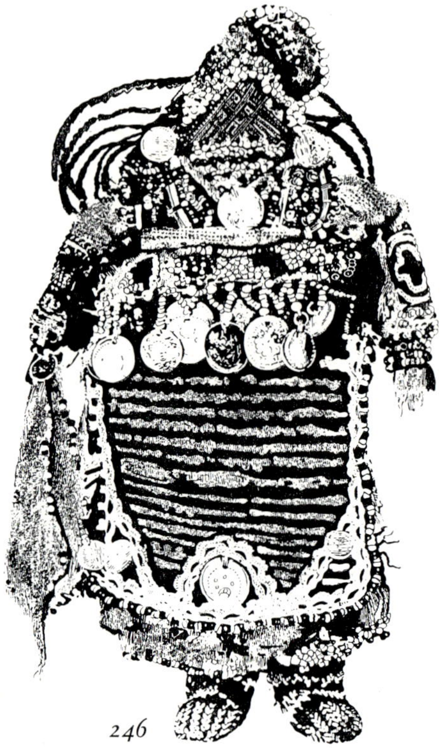

246

Pée, L.: *Kinderspiele,* in *Form,* No. 28 (Opladen, 1964).

Pfeffer, F. (ed.): *Das Puppenbuch* (Berlin, 1921).

Pinon, R.: *Probleme einer europäischen Kinderspielforschung,* in *Hessische Blätter für Volkskunde,* Vol. 58 (Giessen, 1967).

Pollocks World of Toys (catalogue) (London, n.d.).

Popitz, N.: *Die Welt des Kindes* (Munich, 1956).

Poppenhuizen (catalogue) (Amsterdam, 1955).

Portmann, P.: *Die Kinderspiele. Pieter Bruegel d.Ä.* (Berne, 1961).

Puppen von heute, in *Spielzeug von heute,* No. 5 (Sonneberg, 1967).

Puppen im Theater, Film und Fernsehen (catalogue) (Dresden, 1969).

Puppentheater der Welt (by a group of authors) (Berlin, 1965).

Rabecq-Maillard, D. J.: *Histoire du Jouet* (Paris, 1962).

Remise, J., and Fondin, J.: *L'âge d'or des Jouets* (Lausanne, 1967).

Rilke, R. M., and Pritzel, L.: *Puppen* (Munich, 1922).

Roh, J., and Hansmann, C.: *Altes Spielzeug. Auf das Schönste gemacht* (Munich, 1958).

Rumpf, F., and Oswald, E.A.: *Spielzeug der Völker* (Berlin, 1922).

Schaarschmidt-Richter, I.: *Japanische Puppen* (Munich, 1962).

Schickele, R.: *Das Puppenbuch* (Berlin, 1921).

Schmidt, Eva: *Spielzeug und Spiele der Kinder im klassischen Altertum* (Meiningen, 1971).

Schmidt, T., and Mathys, F. K.: *Zur Geschichte und Psychologie des Kinderspiels. (Versuch einer Deutung an Hand eines Gemäldes von Pieter Bruegel d. Ä.)* (Basle, 1964).

Schwedische Volkskunst (catalogue) (Darmstadt, 1966).

Schweizerische Volkskunst (catalogue) (Munich, 1967).

Schwindrazheim, H.: *Altes Spielzeug aus Schleswig-Holstein* (Heide in Holstein, 1957).

Seyfferth, O., and Trier, W.: *Spielzeug* (Berlin, n.d.).

Shpikalov, A.: *Russisches Spielzeug* (Moscow, 1974).

Simmen, R.: *Der mechanische Mensch* (Zurich, 1967).

Sourek, K.: *Volkskunst in Bildern* (Prague, 1956).

Sowjetische Volkskunst (Berlin, 1953).

Spamer, A.: *Weihnachten in alter und neuer Zeit* (Jena, 1937).

Spiel und Spielzeug aus aller Welt (catalogue) (Vienna, 1971/72).

Spiele für Könige und Kinder (catalogue) (Zurich, 1974).

Spielzeug der bürgerlichen Zeit. Zeugnisse für Kindererziehung und Kinderarbeit (catalogue) (Marburg, 1973/74).

Standardisierung (journal) *Spielzeug bzw. Spielzeug von heute* (Sonneberg, 1960 onwards).

Stordeur, Rochus: *Spielzeug in Schleusingen. Informationen über Spielzeug aus Südthüringen* (Sonneberg, 1974).

Swoboda, H.: *Der künstliche Mensch* (Munich, 1961).

247

247 *Doll wearing Algerian costume. Present day*

Thiel, E.: *Geschichte des Kostüms* (Berlin, 1960).
Tischner, H.: *Völkerkunde* (Frankfort-on-Main, 1959).
Vogenauer, E. R.: *Spielzeug* (folder) (Berlin).
Völkerkunde für jedermann (by a group of authors) (Gotha/Leipzig, 1966).
Weber-Wellermann, Ingeborg: *Spielzeugbefragung. Überlegungen anläßlich einer Marburger Ausstellung. Spielzeug als Indikation eines sozialen Systems,* in *Zeitschrift für Volkskunde,* Part II (Stuttgart, 1974).
Wehrhan, K.: *Kinderlied und Kinderspiel* (Leipzig, 1909).
White, G.: *A Bock of Dolls* (London, 1956).
– *Antique Toys and their Background* (London, 1971).
– *Dolls of the World* (London, 1962).
– *European and American Dolls and their Marks and Patents* (London, 1966).
Wilckens, L. von: *Tageslauf im Puppenhaus. Bürgerliches Leben vor dreihundert Jahren* (Munich, 1956).
Wittkop-Ménardeau, G.: *Von Puppen und Marionetten* (Zurich, 1962).
Zabo, C.: *Nürnberger Spielzeug* (Nuremberg, n.d.).
Zingerle, I. V.: *Das deutsche Kinderspiel im Mittelalter* (Innsbruck, 1873).
Zur Geschichte der Kultur und Lebensweise der werktätigen Klassen und Schichten des deutschen Volkes vom 11. Jahrhundert bis 1945 (A short survey by a group of authors.) (Berlin, 1972).